REED'S
SEXTANT
SIMPLIFIED

REED'S SEXTANT SIMPLIFIED

7th EDITION

DAG PIKE

ADLARD COLES NAUTICAL
LONDON

Published by Adlard Coles Nautical
an imprint of A & C Black Publishers Ltd
37 Soho Square, London W1D 3QZ
www.adlardcoles.com

Copyright © Adlard Coles Nautical 1995, 2003

First edition published by Thomas Reed Publications 1995
Reissued by Adlard Coles Nautical 2003

ISBN 0-7136-6705-2

All rights reserved. No part of this publication may be reproduced in any form or by any means - graphic, electronic or mechanical, including photocopying, recording, taping or information storage and retrieval systems - without the prior permission in writing of the publishers.

A CIP catalogue record for this book is available from the British Library.

A & C Black uses paper produced with elemental chlorine-free pulp, harvested from managed sustainable forests.

Printed and bound in Great Britain by The Cromwell Press, Trowbridge, Wilts

Note: While all reasonable care has been taken in the publication of this book, the publisher takes no responsibility for the use of the methods or products described in the book.

CONTENTS

Introduction	vi
Chapter One *Description of the Sextant*	1
Chapter Two *The Principle of the Sextant*	19
Chapter Three *How to Take Sights*	27
Chapter Four *Practical Hints on Taking Sights*	41
Chapter Five *How to Read the Sextant*	59
Chapter Six *Errors and Adjustments*	67
Chapter Seven *Sextant Telescopes and Other Accessories*	82
Chapter Eight *Practical Notes on the Care of the Sextant*	91
Chapter Nine *The Sextant and Coastal Navigation*	97
Index	102

INTRODUCTION

The Sextant In The Electronic Age

The sextant is the purest of navigation instruments. It allows a position line to be determined without reference to any other instruments or systems. With horizontal angles the position can be fixed from marks without any corrections being necessary. Vital ranges can be determined by using vertical angles. Latitude can be found from angles of the sun on the meridian or from the Pole Star. Bring in the chronometer and the position can be determined with adequate accuracy for ocean navigation.

No wonder that the sextant has been the treasured navigation device of generations of seamen and airmen. Its early attraction was the fact that position could be determined with good accuracy for ocean navigation, and it was largely the combination of the sextant and the chronometer which opened up the era of safe and reliable ocean navigation across the Atlantic and Pacific Oceans. Position could be determined without reference to any external systems, although the requirement for clear skies and a co-operative horizon could limit the availability of positions.

It is the independent nature of the sextant/chronometer combination which still makes it attractive in the electronic era. All electronic systems are dependent on external radio transmissions of one sort or another to allow position to be determined. These modern electronic systems can produce accurate positions virtually at the touch of a button, but seamen in particular are concerned not only with accuracy but also with reliability. The time may come when the reliability of the electronic system is such that the sextant becomes redundant. That time is not now, and the sextant is still treasured by a multitude of seamen because of the independence it offers to the navigator.

For many navigators the sextant may have been relegated to the role of a back-up system but that is no excuse for not understanding

Introduction

the way the sextant works and how to get the best out of it. The sextant still has a treasured place in the navigation equipment of many ocean-ranging yachtsmen. It is the traditional way of fixing position, and to the navigator who gets his power from the wind, tradition can be important. For professional and amateur navigator alike, the sextant is still the purest of navigation instruments, and here we explain its design and workings, its accuracy and its care, in the hope that the reverence accorded to the sextant in the past will live long in to the future.

Dag Pike

CHAPTER ONE

Description Of The Sextant

Despite its beauty and apparent complexity the sextant is purely and simply an instrument for measuring angles. It is still carried on all ships throughout the world and by many yachtsmen, and by its aid the navigator may find his way across the oceans anywhere out of sight of land. The instrument is adapted for observing angles in any plane or level and thus vertical, horizontal or diagonal angles may be measured by it. The vertical angles are used for celestial navigation and both horizontal and vertical angles for coastal navigation. Diagonal angles are mainly used for certain aspects of checking the sextant.

The sextant is particularly suitable for use at sea in deep water navigation for measuring the altitude (that is, the angular distance above the horizon) of heavenly bodies above the horizon. Because it is a handheld instrument, accurate measurements of the required angle can be obtained even though the vessel is rolling and pitching. This continual movement of the vessel at sea precludes the use of fixed angle measuring instruments such as theodolites which require a solid platform, hence the importance of the sextant for marine navigation. Because of the Laws of Optics relating to the reflection of light from two mirrors, the angle between two objects can be measured with great accuracy using the sextant. Both of the objects between which the angle is being measured can be brought into the field of view of the sextant telescope simultaneously which means that the movement of the handheld sextant is not critical to the accuracy of the measurement.

A navigator usually possesses his own sextant as one of the most important tools of his trade. All sextants, even the cheaper plastic ones, are kept in a dedicated case when not in use to give them the protection which a precision instrument deserves. Whilst sextants are robustly constructed and can stand up to hard use, it must be appreciated that the sextant is a precision instrument which will only produce accurate

Chapter One

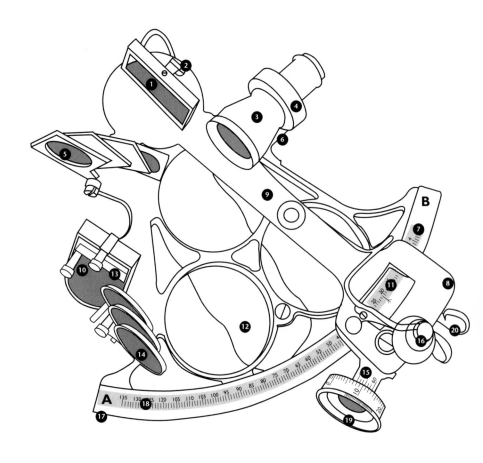

A TYPICAL MICROMETER SEXTANT AND ITS PARTS

1. Index glass
2. Index glass screw
3. Star telescope
4. Telescope collar
5. Index glass shades
6. Rising piece (underneath)
7. Arc of excess
8. Vernier (if provided)
9. Index bar
10. Horizon glass screws
11. The index
12. Handle
13. Horizon glass
14. Horizon glass shades
15. Seconds vernier
16. Electric Light
17. Limb A – B
18. Arc – proper
19. Micrometer head
20. Quick release clamp

measurements if it is treated with care. There is even a correct way of lifting the sextant from its box to reduce the chance of accidental knocks and when in use, the sextant must be treated with the respect which a valuable and treasured instrument deserves. Unknown errors in the sextant caused through mis-handling could be life-threatening if they result in an unreliable position so treat the sextant with care.

The Sextant

It should be stated right away that all sextants, whatever their size and price, do fundamentally the same job. The main advantages of the more expensive instruments are that they are larger, are easier to read, they have finer telescopes and various accessories and may be lighter and easier to handle. The main benefits of a more expensive sextant will be found in the way that it is easier to use, it will be less prone to inaccuracy and is more likely to produce good results in the marginal conditions which often prevail when a position is desperately needed.

Whilst the polished hardwood sextant case looks like an extravagance it is primarily designed to protect the sextant during storage or transit. A quality case is more likely to make the owner respect the contents and the internal stowage provides for all the accessories which go with the sextant. The case is usually fitted with a carrying handle and a lock and key and storage for the key when not in use frequently provided inside the box.

The Sextant Certificate

Fitted inside the lid of the case will be seen the sextant certificate. In the past, all British sextants were sent to the National Physical Laboratory for independent testing. If satisfactory the NPL would then issue a certificate, either Class A or Class B, according to the standard of the instrument and this certificate was permanently fitted in the sextant case. Nowadays it is customary for the certificate to be issued by the makers of the instrument and unless there is a special requirement for official certification this is now the general practice worldwide.

The certificate will give the maker's name and the identification

Chapter One

number of the sextant. This latter is important, because if the instrument is sent for repairs at any time it can be readily traced by this number, together with the maker's name. The certificate sometimes states the radius of the sextant, also the range of the sextant scale and the magnification of the telescopes. A table is provided showing the permanent errors of that particular sextant at various angles, which the user should apply to get the best accuracy. These errors are generally very small and help to give an indication of the quality of the sextant. The date of manufacture is also stated. The sextant certificate is like a guarantee of accuracy and quality and provides the user with a degree of re-assurance.

Handling The Sextant

The main cause of any inaccuracies which may develop in a sextant will come from accidental knocks which can occur even when the sextant is in its case. It is easy to appreciate how the sextant frame or fittings could be knocked just slightly out of true and to prevent the sextant being jarred through being fitted loosely in the case, it is usually firmly held in position by an arm or catch of some sort securing the handle. Foam padding may also be used to cushion the sextant in its case. Before attempting to remove any sextant from its case an examination should be made of the exact arrangement of its fitting and securing so that it is not removed with undue strain.

Having released any securing clips with the right hand, the sextant should be grasped firmly by the fingers around the frame work about its point of balance and lifted from the case by the left hand. As the sextant is raised up it can be transferred to the right hand, which grasps the handle at the back of the instrument. Thereafter the instrument is held in the right hand with all adjustments to the parts of the sextant being made with the left hand.

If the sextant is not being used on a temporary basis it can be laid down on a flat surface. This is a reversal of the removal procedure, setting it down by grasping the framework of the instrument with the left hand, and placing it down on its legs with the handle underneath.

Description Of The Sextant

The greatest care must be taken when lifting the instrument up or placing it down, grasping the sextant firmly by the frame only and never lifting it by the limb or arc, index bar, mirrors, shades telescope or the movable index bar. Holding the sextant at any of these points could cause them to be strained, thus affecting the accuracy of the instrument, particularly over a period of time.

Before putting the sextant back into its case it is essential to fold in the index shades back in towards the middle of the instrument. If they are left in the position for use the sextant will not fit into the case and the shades may be knocked and damaged. The case should be designed so that it is large enough to accommodate the star telescope in position on the sextant although it may be necessary to push in the focusing section. On some sextants it is necessary to remove the telescope altogether which means that the sextant has to be set up each time it is used. When the sextant is in the box and the securing clip fixed lower the lid gently to see if it will close properly but do not force it. If the lid does not close smoothly, open it up again to check that the sextant is seating properly in its stowage and that the mirrors and telescopes do not obstruct the lid.

The case should be large enough to take the instrument at whatever position on the arc at which the movable index bar may be left. This is important, because it enables the sextant reading to be checked subsequently, if any error is thought to have been made in the original reading. However, with many sextant boxes it is necessary to move the index bar some way along the arc to allow the case to close. Usually this is about halfway along the arc.

The Accessories In The Case

The number of accessories which come with the sextant will naturally depend on the quality of the instrument. Older sextants were usually provided with a blank sighting tube and two telescopes, one an erect telescope and the other an inverting telescope, all of which will be provided with a proper housing in the sextant case.

The blank tube is a relatively rare accessory these days. It has no lenses and is merely a sight vane which is used only when taking shore

Chapter One

angles, to keep the line of sight from the eye parallel to the plane of the instrument. An erect telescope or a binocular sight is more likely to be used particularly when the sextant is being used for survey work.

The erect telescopes show objects in their natural relationships but magnified. These are the easiest telescopes to use and are provided to make objects such as stars and the horizon more distinct. With the aid of this type of telescope it is much easier to get a clear picture of the contact between the star and the horizon. Because of their use with star sights, these telescopes are more frequently called a star telescope. Those supplied with modern sextants are bell-shaped and give a much larger field of view than any other type making it easier to locate the star and to distinguish the horizon in marginal conditions. On a sextant where only a single telescope is supplied, this is usually the type used.

The inverting telescope which is longer, shows the object upside down or inverted. It is chiefly used for observation of the sun and has a greater magnifying power than the erect type of telescope. It can be fitted with alternative eye pieces, one with higher magnifying power than the other but the higher the magnification the more limited the field of view. The higher powered eyepiece is preferable when observations are taken on land whilst the smaller powered unit is better suited for sights at sea. Use and experience will soon accustom the observer to the inverted display but this type of telescope is disappearing from modern sextants as improved optics allow a single erect telescope to be used for all requirements. Chapter 7 looks at sextant telescopes in more detail.

Telescope Shades

One or two shades of different densities may be supplied to fit to the eye piece of the telescopes. These are distinct from the shades fitted to the sextant itself and are used when it is required to shade equally the direct and reflected image of the object. They are used for taking sights by the artificial horizon on a clear day or they could also be used if for any reason the index or horizon shades are not in use or have become damaged.

A fully equipped sextant will have most of these accessories and also included in the case may be a screw driver, bottle of oil, camel hair brush, spanner, etc., and a small ball headed pin lever. These are the tools used in the maintenance and adjustment of the sextant and keeping them in the box means they are ready at hand when required. A small piece of chamois leather is usually kept in the case to wipe the mirrors and the arc after use, particularly if there has been salt spray contamination.

The Sextant Itself

The sextant derives its name from the extent of its limb, which is the sixth part of a circle, or 60 degrees. Despite its name, the sextant can actually measure angles up to 120 degrees and the scale is divided accordingly. This is because the rotation of the index bar on which the rotating index mirror is mounted only moves through half the arc being measured. This is explained in more detail later in the next chapter.

Types Of Sextant

In addition to the standard types of marine sextant used on ships for measuring the angles between heavenly bodies and the horizon there are specialised sextants such as those used for survey work. First we will look at the sextants used for star and sun sights and there are three distinct types of these sextants. The differences mainly relate to the fixing of the index bar and the way in which the reading of the sextant angles is done.

These three types are:

a) **The clamping screw sextant:** This type of sextant is now largely a museum piece. This was the basis of the vernier sextant where a clamp secures the index arm to the arc. The broad adjustment of the arm is done with the clamp slack and once it is in the approximate position, the clamp is tightened. This holds the index bar firmly against the arc and when this has been done the index bar may be moved for a short distance along the arc by means of the tangent screw so that more exact readings may be made.

The exact reading of the angle observed is performed by a small arc,

Chapter One

adjacent and butting on to the main arc called the vernier. In this type of sextant it will be found that the vernier can only be moved a little way along the arc to the limit of the tangent screw. This can be very inconvenient in practice because frequently when observing altitudes, the screw will suddenly get to the end of the thread, when it is necessary to unclamp, turn the screw back to the middle of the thread, and resume observation, by which time the sun has possibly disappeared behind the clouds for good.

b) The endless tangent screw sextant. This type of sextant was an improvement on the one described above because by pressing the quick release clamp at the foot of the index bar and holding it open, the bar can be moved by hand anywhere along the arc and automatically clamped by releasing the finger pressure from the quick release.

When this has been done, the tangent screw can be brought into play and as this is endless the index bar can be screwed along from one end of the arc to the other without interruption or requiring further clamping. The clamping screw has been eliminated altogether in this model and replaced by a quick release clamp. The endless tangent screw sextant however, has still to be read by a vernier and whilst these two types of sextant will work and produce good results, they have been almost entirely replaced by the micrometer sextant.

c) The micrometer sextant. This is the most modern type of instrument in which the index bar is first of all fixed at the required position by a quick release clamp. In this way it works like the endless tangent screw sextant but it is in the way that the reading on the arc is measured that it differs. Instead of the vernier adjacent to the arc, a micrometer is used and this enables the reading to be made quickly and accurately. The tangent screw is fitted with a large micrometer head on which the readings are very clearly marked in black on a white ground. This allows a very rapid and easy reading, which is a great advantage, the graduations being easily discernible at arm's length in good daylight, and much more easily than usual in poor daylight, or in dim artificial light. The whole degrees are read direct on the arc and the minutes are read on the micrometer head at sight, thus dispensing with the arc vernier and its necessary magnifying glass entirely. A small vernier is generally provided on the micrometer wheel to enable the angle to be read to 10 seconds (10") when high accuracy is required. For the micrometer to work accurately, the thread in which the micrometer screw works has to be cut with great accuracy. With a vernier sextant it is only the scale which has to be cut accurately.

The Parts Of The Sextant

■ **The frame:** Each sextant manufacturer tends to have their own individual design of frame for their sextants. In designing the frame, the aim is to combine the maximum rigidity and strength with reasonably light weight. Popular designs of frame include the three circle and the diamond frame types. Because the sextant may have to be held for long periods a light frame can be advantageous but if it is too light then it may lose some of its rigidity. A heavier instrument is often favoured as it can give greater steadiness when taking observations in a stiff breeze. The

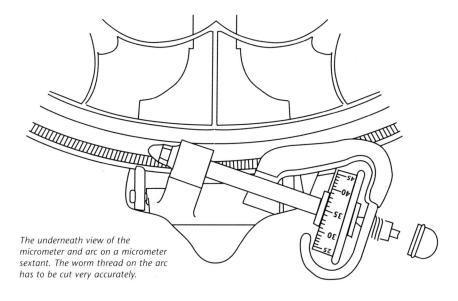

The underneath view of the micrometer and arc on a micrometer sextant. The worm thread on the arc has to be cut very accurately.

Chapter One

average weight is about 4 lbs. although modern plastic sextants are lighter and the specialised survey sextants are made lightweight as they have to be used for considerable periods. Originally sextant frames were made from brass, but aluminium or aluminium alloys are now popular as they are easy to cast, they are corrosion resistant and they are lighter in weight.

At the back of the frame is the fixed wooden or plastic handle of the instrument which enables it to be held in any position between the vertical and the horizontal. The handle is important, as it must be able to be grasped, possibly for longish periods easily and without fatigue. The sextant should be held only by the handle or by the frame and there are three legs one of which may be the handle so that the instrument can be set down safely on table top or other flat surface.

■ **The limb or arc:** The limb of the sextant is often called the arc and is the lower margin or the whole of the curved circular portion of the sextant. The arc is actually the graduated section divided up into degrees which is bedded into the limb. The arc is usually made from a thin piece of silver or platinum, metals with a low co-efficient of expansion which is let in flush with the face of the sextant limb.

The arc proper is graduated in degrees from the right to left, from 0 to 120 degrees although on some sextants, particularly survey sextants, the arc will extend further. To the right of the zero on the arc the scale is graduated for a further 5 degrees and any measurements here are said to be "off the arc". Readings to the left or right of 0 degrees are therefore distinguished as being "on" or "off" the arc respectively. The off the arc readings are generally used for checking the accuracy of the sextant.

In the case of the vernier sextant, each degree on the arc of the sextant is divided into six divisions of 10' each although a few older instruments are still in use that are divided into 4 divisions of 15' each. In micrometer sextants however, each whole degree is cut boldly on the arc with each 10 degrees being numbered and each 5 degrees being indicated by a longer cut than those marking each degree. Some sextant manufacturers mark each 5 degrees and the whole object of this marking is to make it easy to read the scale at a glance. One revolution of the micrometer wheel moves the index along the arc one whole degree, and sub divides the arc to 10" intervals. With a micrometer sextant it is vitally important for the worm drive thread machined into the limb to be very accurate because this is what determines the overall accuracy of the sextant. With a vernier sextant it is the scale which determines the accuracy.

■ **Horizon glass:** This is rigidly fixed to the frame perpendicularly to the plane of the instrument and parallel to the index glass, when the index is at zero. The horizon glass is fixed in a metal frame and the half farthest away from the instrument frame is plain glass and therefore transparent, so that objects may be seen directly through it by the telescope. The other, inner half is a silvered reflector or mirror. This mirror receives the rays of the object reflected from the index glass and transmits them to the observer, viewing through the telescope. It is fitted with two small capped adjusting screws which allow the mirror to be moved in relation to its frame and these provide the means of adjustment to enable the mirror to be made parallel to the index mirror and at right angles to the frame. Springs in the mirror frame hold the mirror against these adjusting screws.

Sextant mirrors can be susceptible to salt spray and on some sextants the horizon mirror is hermetically sealed to make it impervious to salt water or spray. On some types of sextant this may be offered as an optional extra. Another modification from standard is the removal of the top bar of the horizon mirror which gives a clearer view and also gives easier access to the horizon glass silver edge to clean away the salt moisture, which is deposited by the sea air. Whilst most horizon glasses are rectangular in shape some manufacturers offer a circular glass and the choice is really one of personal preference although many users think that the circular mirror produces a clearer image.

■ **Horizon shades:** Three horizon shades of coloured glass of different densities are fitted outside the horizon mirror. These can be turned up as required in order to reduce the excessive glare coming through the

Chapter One

transparent part of the horizon glass. These will be used mainly for sun sights, particularly when the sun is low down and the horizon reflections can be very bright. An alternative to having a series of different filters is to have a single polarising filter which can be rotated to vary the strength of filter. These filters can produce a very sharp horizon line. One of the horizon shades is usually made very dark for use when observing the sun to check index error.

■ **Index Bar (or radius):** The index bar is really a movable arm, being free to rotate on a central axis under the index glass, around a point on the frame which is the centre of the circle of which the arc is a section. It is mounted on a round baseplate beneath the index glass, which is firmly fixed to it at the top of the bar whilst the bottom of the index bar slides along the arc on the limb when pushed by hand. The bearing on which the index bar rotates is critical to the accuracy of the instrument and no play can be tolerated in this bearing. On some sextants this bearing has top and bottom supports with the index arm moving between top and bottom parts of the frame. On others the bearing pin is inserted and then secured with a covering plate.

■ **Index glass:** This is the mirror which is set in the frame attached rigidly to the index arm and perpendicular to the plane of the instrument. The centre of this mirror is over the pivot point of the index arm so that this mirror moves with it and changes its direction as the direction of the index bar is changed. This mirror is silvered all over, and is designed to reflect the image of the sun or any other object upon the horizon glass, whence it is reflected to the eye of the observer through the telescope. The index glass is fitted with a single adjusting screw at the back which is used to adjust the angle of the mirror in relation to the plane of the instrument with a spring fitted into the frame to react against the pressure of the screw. Some modern sextants do not have the adjusting screw, the quality of manufacture ensuring that the frame and hence the mirror are precisely upright and square so that adjustment is not needed. The quality of a sextant is often reflected in the quality of its mirrors and

Description Of The Sextant

like the horizon glass hermetically sealed mirrors can be fitted here. Whilst a rectangular mirror is normal, the index mirror can also be round.

■ **Index shades:** Four index shades of coloured glass, each of a different density are usually fitted. These can be swung into the path between the index and horizon mirrors in order to moderate the brightness of the reflected image of the object. When taking sun sights these shades are essential but they can also be useful when shooting the moon in order to get the correct balance between the bright moon and the dark horizon.

■ **The clamp or clamping screw:** When the index of the sextant has been moved approximately to the correct angle, it must be brought into firm contact with the arc by being fastened at the back of the arc by a clamp or clamping screw. In the case of the endless tangent and micrometer sextants this clamping is automatic as on the release of the finger pressure from the quick release clamp the screw automatically engages with the thread on the frame so that the arm can then only move under the control of the screw.

■ **The tangent screw:** Attached at right angles to the index bar is the tangent screw which comes into operation only when the index bar has first been clamped. It is really a "slow motion" screw and by turning it the index bar is carried backwards or forwards along the arc. This enables the observer to make the images of the objects come precisely into contact because, by its use, the index may be moved with far greater control than can be done by the free swing of the index arm.

■ **The vernier:** On vernier sextants the lower end of the index bar has a dividing scale cut on it which slides along close to the arc. This is called the vernier and it divides the arc into 10 major divisions, each representing one minute of arc, and enables the reading of the index on the arc to be made with accuracy. By establishing which of these divisions is in line with a mark on the fixed scale on the arc it is possible

Chapter One

to read the scale to a much higher degree of accuracy than would be possible by interpolation of the arc marks. This highly ingenious method of upgrading the accuracy of the reading has now been largely superseded by the micrometer which makes it much easier to read the scale accurately but even here a vernier scale is applied so that the micrometer reading can be even more accurate.

With a standard vernier sextant the vernier simply sub-divides the 10' intervals cut into the arc into 10" intervals by each minute of arc being sub-divided into six divisions of 10" each. On an alternative type of scale found on some naval sextants each minute of arc is sub-divided into five smaller divisions each representing therefore 12 seconds of arc (12".) Thus the vernier enables the arc to be read to decimals to the nearest 0.2 of a minute, in accordance with the practice of some nautical almanacs of giving the declination to a decimal point of a minute.

■ **Magnifier:** This is a magnifying glass fitted to all vernier sextants so that the divisions of the arc and vernier may more easily be read particularly at dawn or dusk. It is fitted on a movable arm on the index bar which may be swung to examine the required part of the vernier. Sometimes an opaque glass screen is fitted with which to regulate the natural light for reading the vernier.

■ **Electric light:** A small electric bulb may be fitted to make it easier to read the sextant at dawn or dusk when taking star observations. The battery to power the light is usually carried inside the handle and the light is actuated by a press button at the top of the handle. The light itself is on a shaded movable arm which can be swung away from the arc to read the micrometer as required.

The Removable Parts Of The Sextant

The telescopes are a vital part of the instrument, enlarging the object to make accurate observation easier also helping to ensure that the sextant is held in the proper plane for observation. This is done by aligning the image in the correct area in the telescope field of view. The tube or

Description Of The Sextant

"draw" of the telescope slides in and out to help focus the view and also to adjust the focus for different eyesight characteristics. Many recent sextants use a turning adjustment for focusing similar to that found on binoculars and it is now common to find the telescope fitted with a rubber ring to soften the meeting point between sextant and the eye. The various telescopes which can be used with sextants are looked at in more detail in chapter 7.

■ **Telescope collar:** The telescope collar is provided so that the sextant telescope may be screwed into it and thus kept rigidly in the correct position. The telescope collar is really a double brass ring fitted inside a collar and is so constructed as to furnish means of adjusting the line of sight of the telescope parallel to the plane of the instrument, which in some sextants is just pushed into a socket and kept in position by a thumb screw, and in others is screwed into the rising piece, but is removable at will. On other designs the 'collar' is integral with the telescope, the whole piece fitting into an adjustable slide clamp on the frame. By being adjustable, the line of sight through the horizon mirror can be set. This helps to equalise the illumination of the reflected and the direct images in the mirror and the adjustment may be a milled screw or the slide arrangement. The screw adjustment is more likely when a separate collar is used and here the telescope is simply screwed into the collar.

The normal position of the telescope is where equal parts of the plain and silvered parts of the horizon glass are visible, but by raising or lowering the rising piece the telescope in its collar can be directed to either the silvered or plain glass portions of the horizon glass, and the brilliance of the reflected image be therefore regulated. As the telescope is raised, less of the silvered part of the horizon mirror appears in the field of view and the reflected image will not be so bright. If the telescope is lowered however, the brilliance of the reflected image will be greatest.

■ **Adjusting screws and clips:** These have already been mentioned in connection with the mirrors and the screws, which when they are

Chapter One

tightened push the mirror against the spring clip. Adjusting screws are provided at the back of the index glass and the horizon glass to admit a slight movement of the mirrors to remove errors of the instrument.

In older instruments two small screws were provided in the telescope collar to correct any collimation error. These screws are operated either by an adjusting pin inserted into a hole in the screw head or by means of a small screwdriver.

The Plastic Sextant

In this modern synthetic age the use of various plastic materials as substitutes for more conventional materials in manufacturing processes, is on an ever increasing scale. It is not, therefore surprising to find that plastics have found their way into sextant manufacture. It is now possible to purchase a useful, workmanlike and effective plastic micrometer sextant. These are of particular interest to yachtsmen because he may be deterred from using the conventional sextant for several reasons, notably price, weight and bulk. On a yacht there is a much greater risk of damaging the sextant when taking sights, either by knocking it against something if the boat lurches suddenly or through salt spray.

With the plastic sextant costing a fraction of the price of the conventional sextant, weighing about 1.1/2 lbs and being a reasonably robust instrument it is eminently suitable for the sort of rugged treatment a sextant might have to endure on a yacht. The low price means that it can be an economical proposition to carry two sextants on board which helps to guard against the risk of accidental damage. Even if a metal sextant is carried in the interests of accuracy, the plastic sextant can be carried as a useful backup.

Obviously such an inexpensive and simple sextant cannot be expected to have the accuracy of a first class precision instrument, but if its limitations are fully appreciated it can be extremely useful and well suited for navigation in small craft. Its use increases the navigational tools available to the coastal navigator; horizontal and vertical angles are easily measured and once accustomed to its use in this way, the navigator may be encouraged to develop his navigational skills still further.

Description Of The Sextant

Plastic sextants follow most of the conventional sextant design trends and they are moulded usually from a high density thermoplastic polycarbonate, which has a relatively high resistance to temperature change although it is more susceptible than the conventional brass. The actual coefficient of expansion of the plastic is around three times that of brass so that larger errors must be expected with such sextants. While this may seem rather alarming, a plastic sextant may be designed for use at an ambient temperature of 70 degrees F, the expansion of contraction of the scale when temperatures change is around 1 minute of arc for each 5 degrees F variation above or below the designed operating temperature. From the average yachtsman's point of view therefore, the degree of error is not as wide as would appear at first sight. This factor however, inevitably introduces a variation in the index error, with variation in ambient temperature. It is essential to check the index error at frequent intervals on a plastic sextant and this is best done immediately after taking a sight. Provided this simple drill is followed and the necessary corrections applied, the accuracy of sights should be within acceptable limits for most navigation requirements provided that the possibility of larger errors is acknowledged and compensated for. A plastic sextant should never be left out in the sun otherwise the temperature change can be quite alarming and distortion may occur.

Apart from the temperature variation the accuracy of any sextant depends on the precision with which certain moving parts are machined. In the plastic sextant the manufacturing process is said to produce an average error of 1 minute on the full arc of the instrument. It is important to appreciate that the accuracy of a fix in a small craft depends on a variety of factors. The difficulty of taking a good sight from a small, unstable platform, probably moving in a violent manner, is such that the navigator can reasonably consider a fix within 3 miles a good result and five miles acceptable. With practice and skill these accuracy levels can be improved on. Whilst it is not considered that instrument errors and errors due to poor sights will cancel out, it is suggested that under average conditions in a sea- going yacht with a relatively inexperienced navigator, results with a plastic sextant could compare very favourably

Chapter One

with a precision instrument. For the professional the level of accuracy of a plastic sextant may introduce too much of an element of doubt but provided that allowances are made for potential inaccuracies when navigating, then useable position information can be obtained from these plastic sextants.

The plastic sextant can fulfil the requirements of economy and they have a place in the repertoire of a navigator but there is no real substitute for a proper sextant. The pleasure of using such a beautifully made instrument and from it working out your position can be considerable. The wonderful feeling of independence which this process gives is hard to contemplate until you have tried it. It is the stuff of true navigation and the sextant is really in its element in blue water sailing. Understanding the component parts of the sextant helps to understand the whole. Loving care has gone into the manufacture of the sextant and the same loving care and understanding will help you get the best out of it.

CHAPTER TWO

The Principle Of The Sextant

Optical Laws

There are two important optical laws which are used in the design of the sextant and which make it work. The construction of the sextant is based around these optical laws and to understand the laws is to understand the sextant and its use and correction. The laws are:-

■ **The Angle Of Incidence Equals The Angle Of Reflection:** If a ray of light strikes a mirrored surface exactly at right angles to the surface then it is reflected back in exactly the same direction. In view of the above law, this is logical because the ray hits the mirror at 90 degrees and is reflected back at the same angle. Now if the angle at which the ray of light hits the mirror is oblique to the mirror's surface, in accordance with the above law, it will be reflected back at the same angle but in the opposite direction, using a line at right angles to the sextant as the 0. In the sextant this law applies to the index mirror where the light from a star, the moon or the sun strikes this mirror at an angle and is reflected down to the horizon mirror. Here the ray of light is reflected and the same theory applies with the light being reflected towards the viewing telescope at exactly the same angle at which it strikes the horizon mirror.

It is this reflection process which forms the basis of the sextant theory and which allows the image of the heavenly body to be brought down to the horizon mirror and then to the telescope so that the heavenly body and the horizon can be lined up and the angle between them measured. The angle at which the ray of light strikes the mirror is called the angle of incidence and the angle at which the ray of light leaves the mirror is called, quite logically, the angle of reflection and these two angles will always be equal.

Chapter Two

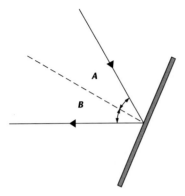

The angle of incidence equals the angle of reflection. Angle A will equal angle B at whatever angle the initial ray strikes the mirror.

■ **If a mirror is rotated through an angle, then a ray of light reflected from the mirror moves through twice that angle:** Another way of putting this is to say that the angle between the first and last directions of a ray of light is twice the angle of rotation of the mirror. It is this optical law which works in favour of the sextant designer and enables a sextant to be made into a conveniently sized and easy to handle package. It means that the index arm and hence the index mirror attached to it, only has to move through half the angle being measured so that although the scale on the arc of the sextant measures to over 120 degrees, it only extends over an arc of a little over 60 degrees from its centre point. This is because the index mirror fixed to the arm only has to move through half the angle that the ray of light from the heavenly body has to move. It is quite easy to see how this law works when you bear in mind the first law which states that the angle of incidence equals the angle of reflection. When the mirror is rotated both the angle of incidence and the angle of reflection either increase or decrease by the same amount as the mirror is rotated. This means that as both angles are changed by the same amount as the mirror is rotated the reflected beam moves through twice the angle of the mirror. The sextant is used to measure the angle between the heavenly body and the horizon. The fact that the index mirror and the horizon mirror are in slightly different places would have a parallax effect on the angle being measured, but in practice this effect is negligible because the difference in the positions of the mirrors is insignificant in relation to the distance of the heavenly bodies from the sextant. The effect of the distance apart of the mirrors could be more significant when the sextant is used to measure horizontal angles from shore objects or for vertical sextant angles measured for distance off because the sighted objects are

The Principle Of The Sextant

closer, but in practice, even here the difference does not have a significant effect on the results.

"Taking sights" is the nautical term for observing the altitude of one or more of the heavenly bodies with a sextant in order to ascertain the vessel's position at sea. This may be done several times a day when out of sight of land and it is an important part of a vessel's routine if sextant positions are the only means of fixing position. Indeed, the vital noon sights tend to take priority over everything as they produce the vital latitude element of the fix. The method of taking sights under different conditions will be considered later but as part of this theory of the sextant it can be helpful to understand just what is done and what is achieved when taking sights.

An observer on a vessel at sea, looking out horizontally sees a circle bounding his view, and this apparent meeting place of the sea and the sky is called the sea horizon. This horizon appears to the observer to be flat and level and in clear weather it is visible for the whole 360 degrees of the circle bounding his view. Looking skywards he sees one half of the heavenly world, which is called the Celestial Concave. This extends for 180 degrees, the arc of a circle which extends from any one point of the horizon to the diametrically opposite point.

In the daytime the observer may see the sun or the moon or both together. These can be seen with the naked eye and by using the telescope it may be possible to see the planets Venus and Jupiter. These are the only heavenly bodies which are available for position fixing during the hours of daylight.

During the night time the choice of heavenly bodies available is much wider. There is a wide selection of stars and some or all of the four planets Venus, Jupiter, Mars and Saturn will be visible. The Moon is widely available at night and the only body which can be guaranteed to be missing is the Sun. The other thing which is missing at night is the horizon so that it is only possible to take night sights during the limited periods at dawn and dusk when both the horizon and the heavenly bodies are in view simultaneously.

Having decided which of the available heavenly bodies he is going to

Chapter Two

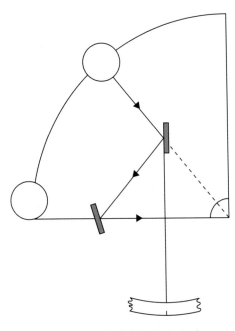

The sextant measures the height of a body above the horizon, but the navigator actually wants the angle between the body and the zenith for most calculations.

use for the sights the navigator now has to use the sextant to measure the angle between the sea horizon, generally called the visible horizon, and the observed body.

Now this angle, taken with a sextant, of a heavenly body above the sea horizon is called its altitude, and because various corrections have to be applied later to this, the actual altitude read from the sextant is known as the sextant altitude. After applying a series of corrections this sextant altitude becomes the observed altitude and this, combined with the precise time of the observation, allow a position line to be calculated. When this is plotted on the chart it will tell the navigator that the vessel lies somewhere along this position line. It needs at least two of these position lines to fix the position and they should be as wide apart in angle as possible so that any error in one position line will have the minimum effect on the accuracy of the position. This means that you need observations from at least two heavenly bodies in order to fix the position and ideally they should be 90 degrees apart in horizontal angle but a 30 degrees horizontal angle will still give a reasonable cross of the position lines. The position can be fixed from a single heavenly body provided sights are taken with a time interval of at least 3 or 4 hours. In this situation, the first position line obtained is carried forward along the course and distance travelled by the vessel so that it can correspond in time with the second position line. Coming back to the sextant, when an

The Principle Of The Sextant

observer places his eye at the end of the sextant telescope, the telescope is pointed towards the horizon in the direction of the heavenly body. The horizon is seen through the unsilvered or clear section of the horizon mirror. If the sextant is set at zero the line of the horizon will also be seen in the silvered part of the mirror. This means that it is the direct image which is seen through the unsilvered part of the horizon glass (or over the top of a mirror which has no unsilvered part) and it is the reflected image which is seen in the silvered section. Now, when the heavenly body being observed is at an elevation of 60 degrees and the sextant index arm is set at this angle the observer should see the body in the silvered section of the horizon mirror and it can then be lined up accurately with the horizon to get an accurate measurement of the elevation by reading the sextant scale. Whilst this sextant scale will read the accurate elevation, the index arm will only have moved half that angle along the scale because by moving the index mirror through 30 degrees, the ray of light from the heavenly body will move through 60 degrees, the angle being measured.

One important point to remember when taking sights is that the actual angle which is required for calculating the position line is that between the heavenly body and the point immediately above the observer. This point above the observer is called the zenith and the angle between the heavenly body and the zenith is called the zenith distance. The problem is that there is nothing in the sky to indicate to the observer where the zenith is actually located. However, as the angle between the zenith and the horizon is exactly 90 degrees, and the horizon is clearly defined, it is easy to measure the angle between the heavenly body and the horizon and subtract this from 90 degrees to obtain the zenith distance.

The actual angle or altitude which is required to be measured by the sextant for calculating position is that from the centre of the body to the horizon. For stars, the pinprick of light which represents the star can be taken as the centre but as the sun, moon, and planets each have a perceptible disc, their centres cannot be observed accurately.

The only clearly defined part of these bodies which can be observed

Chapter Two

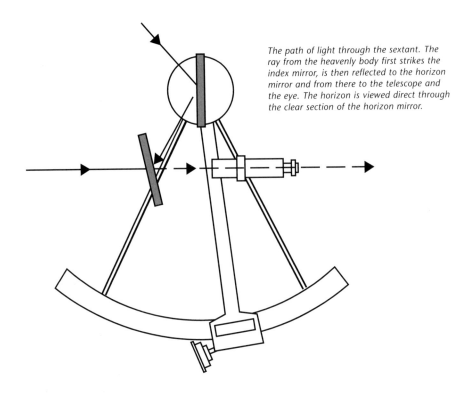

The path of light through the sextant. The ray from the heavenly body first strikes the index mirror, is then reflected to the horizon mirror and from there to the telescope and the eye. The horizon is viewed direct through the clear section of the horizon mirror.

is the circumference and so measurements are taken from the upper or lower, usually the latter, parts of the circumference. This is called the limb of the body, not to be confused with the limb of the sextant so that when an observation is taken of the lower circumference of the sun we say we are observing the sun's lower limb. Corrections are then applied so that the resulting angle corresponds with the centre of the body.

With the sun it is the lower limb which is almost invariably observed because it is easier to line up with the horizon. With the moon the choice of which limb to use is not always open and what is called the enlightened limb is always used. This is the limb which has the full curve of the circumference and is easy to determine when the moon is nearly new, but care has to be taken when the moon is nearly full that the correct limb of the moon is used as it can present a barely distorted circle.

For most practical purposes when taking sights at sea the diameter of a planet can be ignored. Rather than apply any correction the observer can line the horizon with the perceived centre of the planet which will give adequate accuracy for practical position fixing.

Owing to the earth's daily revolution all the heavenly bodies appear to rise in the east and slowly increase in altitude until a certain point is reached. They then decrease in altitude until they set in the west. This fundamental movement is going on all the time though clouds may obscure the bodies, and daylight of course makes it impossible to see the stars at this time.

If an observer faces true north or south at sea, then an imaginary line running from the true North Pole to the true South Pole passing directly overhead through his zenith is called a meridian, in this case being known as the observer's meridian. Now this imaginary line marks the point when the altitude of a heavenly body ceases to rise and commences to fall, or in other words, when it ceases to bear easterly at all, but bears exactly north or south of an observer, before starting to bear westerly.

Now when a heavenly body reaches the observer's meridian, it is said to culminate, and thus at culmination attains its greatest altitude. The time of culmination is called the time of transit or time of meridian passage which is the time the body will cross the observer's meridian. Taking a sight at this time is valuable for the navigator because it can produce a position line without the need for a time reference. The altitude of the body is observed as it increases and then levels off before starting to decline. The highest altitude reached is used to calculate the position line, which because the body will always be north or south at this point will create a position line running east and west, hence indicating latitude.

From the foregoing it will be seen that there are two distinct types of "sights", one taken when the body is moving, ie; rising or falling, and the other taken when the body is stationary in elevation. The former is called an observed altitude and the latter is called a meridian altitude which occurs whenever a heavenly body crosses the observer's meridian, hence the name. The speed of the body's movement and the length of time it is stationary on the meridian depends entirely on the relative positions of the observer and the body. For example, when the declination of the body is about the same as the observer's latitude, the sun has a long way to climb, because it will be nearly overhead with a meridian altitude of about 90 degrees at transit. It will thus appear to rise and fall rapidly, and will remain stationary at transit for a matter of seconds only.

On the other hand, if the observer is in, say 60 degrees north latitude, whilst the sun is on the equator, the meridian altitude at transit will be about 30 degrees only, so the rise and fall will appear to be very slow and the sun will appear to "stand still" at transit for several moments. In these cases the exact time of transit may not be able to be observed effectively, so "time" is given a secondary place to observing the highest altitude, which is the meridian altitude.

The theory of the sextant is quite simple to understand and it is an extremely effective and practical instrument, capable of being used under difficult conditions. There were a number of ancestors to the sextant, many of them large and unwieldy which were difficult to use on a moving vessel. Before the mirror system was developed a plumb line was used to denote the vertical and hence the zenith, but the sextant and its principle of double reflection enabled a compact, and relatively lightweight and easy to handle instrument to be developed, which is also capable of measuring angles with great accuracy when used in competent hands.

CHAPTER THREE

How To Take Sights

Taking sights of heavenly bodies is largely a question of technique or skill. Practice is the best way to attain the skills and there is no real substitute for this. A skilled sextant user will be able to snatch a sextant sight on a brief appearance of the sun or in near impossible conditions at dawn or dusk. Whilst practice is essential so is a grounding in the basic sextant skills and this is what this chapter is all about.

In the last chapter we saw that there were two different types of sight, the observed altitude and the meridian altitude. Each of these requires a different technique and organisation in order to get the best results. One of the first noticeable differences between the two is that observed altitudes can be taken when a suitable body and an horizon are available and the timing can largely suit the convenience of the observer. For a meridian altitude the timing is dictated by the time the body crosses the meridian. There is no negotiation about this and so meridian altitudes are time critical. For the first time user of the sextant, observed altitudes are much easier for practice because they can be taken at any time during the day when clouds permit so that you can quickly become familiar with the use of the sextant.

The beginner should practise continually bringing the sun down to the horizon, and initially the telescope may be dispensed with, just look through the telescope collar, so that there is a much bigger field of view to help keep the sun in sight whilst it is brought down to the horizon. Practice on land is a good starting point in order to get the feel of the sextant. Here you will not have to cope with the inevitable movement of the vessel which can make life difficult for the beginner. A position on a beach is ideal but it is not difficult to contrive a suitable substitute horizon away from the sea such as a long wall or a rooftop, just to get used to handling the sextant.

Chapter Three

Finding a suitable position on board a vessel can be more difficult. You want a position where the vessel is steadiest which usually means somewhere amidships but equally important is that the position allows the sextant user to stand or brace himself comfortably and be protected from spray or rain. On board a ship the sights will almost certainly be taken from the bridge provided that there is a clear view of the horizon in the required direction. If the sun is on the weather side then it is often possible to get shelter behind a deck house. You will almost certainly be high enough to get a clear view of the horizon without the waves rearing up between you and the horizon to create a false horizon.

On a yacht or small craft, finding a suitable position for taking a sight can be much more difficult. The cockpit is probably the safest place and here you can usually brace against the side of the cockpit to steady yourself and there should be a reasonable degree of protection particularly if a dodger is fitted. The main snag with this position is that you are very low down, only a few feet above sea level and unless it is a relatively calm day you will find it difficult to get a true horizon. Standing on the coach roof creates probably the highest point on the yacht for taking sights but here there is little to brace against unless you kneel down or lean against the mast and usually there is little to give you a secure grip against the motion of the boat. Probably the best compromise position is at the shrouds where you can stand up and partially secure yourself by leaning against the shrouds with the sextant arm wrapped round the shrouds as added security. Holding onto the mast is another alternative which also has the benefit of a higher height of eye to give a clearer horizon. Neither is ideal and these positions are not really recommended in rough weather. So much will depend on the boat and the conditions but do bear in mind that you can alter course to get a better motion for the boat or a better view of the sun or stars. Always wear a safety harness well secured during sight taking and this safety harness can be used to lean back against and with your feet wide apart create a sort of triangulated secure pose for sight taking. Using a safety harness can delay the time you take to get to the cockpit for a chronometer reading and you have to find a balance between security and convenience. It can be a good idea to have the sextant on a strap placed around your neck so that even if you do have to make a grab if there is a sudden lurch at least you won't lose the sextant and another practical tip is to have a pillow or cushion placed in the cockpit so that you have somewhere to put the sextant down when it can immediately be secure.

Taking sights in rough seas from a yacht requires a considerable degree of skill and practice to get meaningful results. Before getting the sextant out get a feel for the way the yacht is rolling or pitching. What you want is a slack period when the yacht is steady for a short period of time and this usually occurs at the end of each roll unless the yacht is very stiff and tends to snap back. Ideally you want this momentary period of stability to coincide with the yacht being on the top of a wave. In rough seas there is little point in taking sights at any other time than when the yacht is at the top of a wave because you will not get a true horizon from any other point. Only when you have got this feel for the movement of the boat and worked out where you will take the sight from should you get the sextant out. Under these conditions you will need to take a series of sights and experience will tell you the ones which are good. In assessing sights you are looking for those taken when the yacht was reasonably steady, when it was near or on the top of a wave and when you had the sun 'kissing' the horizon. Getting all three of these requirements at the same time is difficult and you must be discriminating about the quality of your sights, remembering that a bad sight will give you a bad position. You will have a second chance to see any obviously bad sights when you come to plot the position lines so that they can be removed.

When it is time to get the sextant out it is removed carefully from its case with the left hand gripping the metal frame of the sextant. Once clear, grasp the handle with the right hand. The sextant does not like being knocked so always move with care when you have the sextant in your hand and on small craft it must be one hand for the sextant and one to hold on with which does mean that you have to move carefully from handhold to handhold. You will need to time your movements with extra

Chapter Three

care because you are handicapped with one hand occupied with the sextant. Before coming out on deck it can be a good idea to set the shades to what you think you will need and if you know it you can set a rough altitude. The telescope can also be set and focused if this is necessary. It can save a bit of work and time when you are out in the exposed sighting position and reduce your exposure to risk.

For taking the sight you hold the sextant vertically in the right hand. Unless you know the rough altitude set the index arm at zero. Before taking any observations of the sun the shades must be used to eliminate glare. When observing the sun it is **always** necessary to use the index shades, otherwise the sun's rays, on being reflected back to the observer's eye, produce such a glare that it is impossible to see the sun at all and you can damage your eyes. These shades may be used singly or in combination, and when the sun is very brilliant, two shades are often necessary. Generally it is best to turn in the lightest coloured index shade to commence with, altering this to a "deeper" shade if the sun seems too strong.

When the sextant is pointed directly at the sun, it may be necessary to also cover the horizon glass by a shade to reduce the reflected glare off the sea. This shade will also make the horizon line much clearer. Use the darkest shade first to reduce the chance of eye damage and this should be a general rule with all shade use.

Now, holding the sextant vertically and with the index set at zero, point the telescope collar and horizon glass directly at the sun. It will be obvious at once whether the correct shades have been used, by the sun's face being clearly defined as a bright coloured circle, neither too strong or too weak. If however, the glare makes it impossible to see the sun properly, then turn up "stronger" shades, and again direct the sextant at the sun.

We now see the actual or true sun through the unsilvered half of the horizon glass, whilst coincident with it or superimposed upon it but not visible because the sextant is in adjustment, is the sun's image reflected to the eye from the index glass via the silvered portion of the horizon glass. The clamping screw should now be loosened sufficiently or the quick release clamp pressed, and kept pressed, to allow the index bar to slide easily along the arc. Like most things associated with the sextant it is important to do this correctly. With the clamping screw sextant the index bar should be held firmly between the left thumb, (nearest the body) and the second (longest) finger resting on the tangent screw. The first or index finger should be laid out in a natural manner ahead of the index bar touching the bottom of the limb so that as the index bar is pushed along the arc, the index finger is "pushed ahead" thus retarding the index bar and giving to it a uniform and constant movement.

With the automatic clamping sextant the quick release clamp must be released from the thread by squeezing two projecting parts of the quick release together. Whilst most observers use the left thumb and index finger to do this, with practice this type of sextant can be manipulated exactly as described above for the clamping screw model to give a more controlled movement.

Now, keeping the eye fixed on the sun, by applying gentle pressure to the index bar, the true and reflected suns will separate rapidly. This of course does not matter, as all we are concerned about is to bring the sun's image down to the horizon.

By this time we shall have to give up looking at the true sun direct, and gradually lower our line of sight (and therefore the angle of the whole sextant). We now follow the sun's image (seen in the reflected part of the horizon mirror), which will descend at the same speed as the observer pushes the index bar along the arc, the pressure being applied slowly and smoothly as described above.

When the image is at or near the sea horizon, which can be seen through the plain part of the horizon glass, the angle on the sextant will be, of course, the approximate observed altitude. The index bar should now be tightened by the clamping screw or by withdrawing the finger and thumb pressure from the quick release clamp when it will automatically become clamped.

Unless the altitude is low, there will probably be little glare on the horizon, so the horizon shades may be dispensed with, and folded back out of the way so the horizon may be seen more easily. These horizon

Chapter Three

shades are most useful at low altitudes after sunrise or before sunset, if the sun's rays are strong, and should be employed until all horizon glare is eliminated without unduly dimming the horizon line.

When the sun has been reflected down to the horizon, each of the index shades should be folded in, in turn, so as to find which is most suitable. The whole object being, of course to get the sun's limb defined clearly but without glare.

When practising with the sextant this operation should be repeated frequently until it can be done quickly and accurately and virtually automatically. This will teach you how to handle the sextant but experienced navigators generally dispense with the above method and simply point the sextant at the horizon as directly as possible underneath the sun. Then having turned in, initially the dense index shade, move the index bar away slowly until the sun is seen in the silvered (reflected) part of the horizon glass near the horizon which is viewed through the plain glass part of the horizon mirror.

At first you may have to move the index bar backwards and forwards several times, until you see the sun, as probably you will find that you have not got the instrument pointed directly underneath the sun. It generally saves time to sweep continuously round the horizon a little to the right and left of the estimated descending path of the sun to avoid missing it.

Now, as we are preparing to take a sight properly, we must insert the telescope. The ordinary erect or star telescope should be used although many modern telescopes may only have one general purpose telescope. When you change or fit a telescope the most important point to be careful about is not to burr the thread when screwing the telescope into the collar. It will generally be found most satisfactory when doing this to hold the sextant with the arc vertical so that the weight of the telescope is downwards and will thus help the thread to engage correctly.

Modern sextant telescopes usually have interrupted threads which are fixed in the collar simply by a half turn of the telescope in the thread, a very convenient arrangement. There is generally a small portion of the rim of the telescope cut away to form a flat surface to act as a guide to insert the interrupted thread in the correct position. Other types may have the vertical clamping bar fixed permanently to the telescope view a vertical slide adjustment incorporated into the clamping bar for height adjustment.

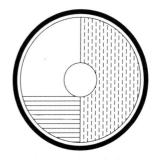

Bringing the sun down to the horizon as viewed through the horizon mirror. Here the lower limb of the sun is being used.

Having fixed the telescope, which should be left in the sextant thereafter and put away in the sextant case in this position, the eye piece should be pulled out until the telescope is focused to suit the eye of the observer. It is best to focus it on a distant vessel, land, or the farthest point away in the ship.

When using the telescope, the eye not being used for sighting the object is generally closed although some observers always keep both eyes open. If the blank tube is being used however, both eyes should be used, two eyes being better than one when they are both being used under the same magnitude. Adjust the index shades to tone the sun's reflection down so it may be seen clearly without glare, and turn up a horizon shade if there is a glare on the horizon. Generally the horizon shades are unnecessary when the sun is high in the heavens. Now direct the sextant straight at the horizon beneath the sun, and move the index bar along the arc until you see the sun just near the horizon as already described. Clamp the index bar.

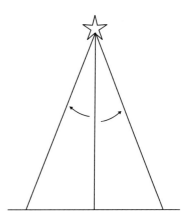

When taking an observation the sextant has to be vertical. This diagram demonstrates how a larger angle will be measured if the sextant is not exactly vertical.

As explained in Chapter 2, it is customary to observe the sun's lower limb on the horizon. Having got the approximate altitude clamped on

Chapter Three

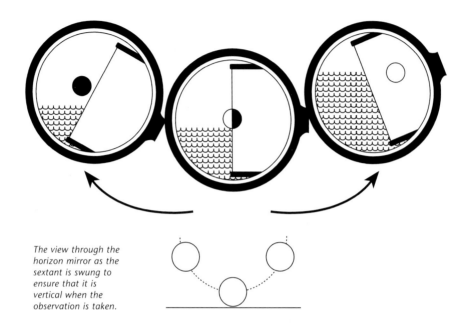

The view through the horizon mirror as the sextant is swung to ensure that it is vertical when the observation is taken.

the sextant as described above, exact contact is made by means of the 'slow motion' screw (called the tangent or micrometertangent screw). When the bottom of the sun just touches the horizon it is termed 'kissing' the horizon.

We know that when taking an observed altitude the body is rising or falling, so it can only kiss the horizon momentarily. In fact, the observer really has to turn his tangent screw at the same speed (but in the opposite direction) as the movement of the body, so as to make certain that the sun's lower limb is just touching the horizon when the observation is made and recorded. It is important also that the reflected sun be brought down exactly underneath the actual sun and the angle is measured vertically, otherwise the altitude measured will be higher than it should be. Holding the sextant exactly vertical when taking a sight is a skill and the correct altitude is obtained by swinging or rocking the sextant to and from about a point somewhere near the index mirror in a pendulum fashion so that the vertical point can be seen as the reflected sun moves along the horizon. Some sextants have a small spirit level attached to the horizon mirror to check that the sextant is upright but this is not a very practical solution at sea.

Only constant practice will give completely reliable results. When making the final adjustments the sextant should be held using the thumb and fingers of the left hand to turn the tangent screw or micrometer wheel slowly to counteract the sun's movement. It should be observed that the tangent screw of the vernier sextant does not act until the index bar is fixed by the clamping screw at the back of the lower part of the index bar. Care should be taken not to force the tangent screw when it reaches either extremity of its thread.

When the index bar is to be moved a considerable distance along the arc on a vernier sextant, the clamping screw must first be loosened. When the index is brought approximately to the division required, the clamping screw should be lightly tightened and exact contact made with the index by gradual movement of the tangent screw.

Having made a good contact between the sun and the horizon by means of the tangent or micrometer screw adjustment, whilst at the same time rotating the sextant gently pendulum fashion so that the observer knows he has taken the correct vertical angle, this is the moment to record the time. This time is just as vital to getting an accurate position as the altitude and the ideal way is to call "stop" to another observer who is stationed at the chronometer who then takes the exact time when "stop" was called.

Needless to say, the observer must not move the screw again after calling stop but must read the altitude on the sextant. This can wait until the observer is back in the cockpit on a yacht but on a ship it is read off and recorded along with the time in the sight notebook ready for calculating later. If the observer feels that the contact between the sun and the horizon was not really very good, he should discard this faulty observation and take another and better observation if possible. It is very important to be honest about the quality of a sight because inaccuracies will reflect directly in the accuracy of the position obtained. However if a sight of the sun has to be snatched because of cloud then such position

Chapter Three

will be better than nothing but treat it with more caution. Experience will let you gauge the quality of a sight.

To Observe An Altitude Of The Moon: The primary requirement for any sight taking is a clearly defined horizon and this usually dictates when moon sights can be taken. Therefore the best time to take a moon sight is during daylight, and at dawn or dusk, though frequent opportunities appear to occur to get observations during the night. Take care with moon sights at night that it is the true horizon that you are using as the moon's reflection which is used to delineate the horizon may not extend all the way to the horizon. Night sights of the moon are best taken when the moon is fairly low although a full moon may generate enough light to provide a useable horizon.

The observation should be made exactly as described for the sun, except that the visible limb of the moon, called the enlightened limb, is always observed. This may be either the lower or the upper limb but this does not matter as long as a note is made in the sight notebook of which limb has been observed.

To Observe The Altitude Of A Star: Stars should always be observed during morning and evening twilight, when the horizon is clearly defined. Although occasionally during the night the horizon can be seen so clearly that star sights may be taken, it is rarely possible to be certain that you can define the horizon exactly, hence for all practical purposes stars may only be used for sights at twilight.

The procedure for observing stars is somewhat different from that adopted for the sun. When taking sights at dusk only the brightest stars will be visible before the horizon becomes too dark to be clearly defined. At dawn however, the stars will have been visible to the observer for some time, so it may have been possible to take the approximate altitude some time previously. In any case, at dawn, provided there is clear visibility, the navigator can more readily decide which stars to use, and can see their position. No sextant mirror shades are required for star work, so they should be folded back out of the way.

How To Take Sights

There are two main ways of observing stars, first, when you can see the star easily and there can be no mistake in its identity, and secondly when the star is invisible until just before the sight is taken as at dusk or when it is ill-defined as in the case of a weak star.

Holding the sextant in the reversed position so that the horizon can be brought up to the sun.

■ **1. When you can see the star easily:** In this case the sextant is clamped at zero and the star telescope inserted and focused. The rising piece should be screwed well down to the frame as this tends to increase the brightness of the image. Now a bright star can be brought down to the horizon in the same way that has been described for the sun. It is much simpler and saves time however, to take the horizon up to the star. This can be especially valuable with a high altitude star, or a second or third magnitude star, or if using an older sextant with small mirrors where it may be difficult to hold the star in the frame when bringing it down to the horizon.

To take the horizon up to the star the sextant is held in the left hand with the index bar set around zero. Held in the left hand the sextant is upside down with the arc at the top and with the end of the arc resting on or close to the forehead. The telescope is pointed straight at the star so that the star is actually seen in the plain portion of the horizon glass. Now, keeping the star fixed in sight in the plain glass all the time, unclamp the index bar with the right hand, and move the index bar away until you see the horizon come into view in the horizon mirror near the star. Clamp the index bar, reverse the sextant to the normal position and then it is a simple matter to make exact contact between the star and the horizon using the tangent or micrometer screw as described for the sun.

Chapter Three

Although this sounds complicated, it is really very easy and can greatly simplify star sights particularly when the time available for a series of sights is limited. It is quick, and as the star is kept in sight the whole time, the mistake cannot be made of using the wrong star. Also, as the horizon is a long continuous black line, it is very easy to see when it arrives up at the star. This same procedure could be adopted for the sun but the brightness and size of the sun makes it unnecessary. A star has no appreciable size so the centre of the star itself is brought to the horizon and made to cut as accurately as possible.

■ **2. When the star is invisible or ill-defined:** When star observations are required at dusk, one may not be able to recognise the smaller stars before darkness falls and renders the horizon too ill-defined for accuracy. In this case the approximate altitude of the star or stars can be calculated in advance, and the sextant set to this angle before the sight is taken so that the star should automatically appear close to the horizon. This can be a valuable procedure if the bright stars do not offer a good cross for position fixing or if there is an urgent need for obtaining good observations it is a very wise precaution to calculate the approximate altitudes in advance, just in case.

Once set with the altitude the sextant is now pointed at the horizon in the approximate direction which can also be calculated. By sweeping round a little to the right and left, the star should be seen near the horizon, when exact contact may be made with the tangent or micrometer screw in the usual way. The point of the horizon at which to look can be found from a star globe, atlas or map or from azimuth tables.

To Observe The Altitude Of A Planet: Observations of planets are dealt with in exactly the same way as with stars, except that as the planets Venus and Jupiter are so bright, it is generally unnecessary to calculate their altitude in advance. The horizon can be reflected up to a planet exactly as described above for a star. To obviate any correction for semi-diameter, the horizon should be made to cut through the centre of the planet which gives adequate accuracy.

How To Take Sights

If it is desired to take the altitude of Venus in the day time, then the approximate altitude must be calculated in advance and clamped on the sextant. Provided there is little or no cloud and Venus is not too close to the sun for it to be overwhelmed by the sun's light then, by sweeping the horizon in the correct direction Venus should be picked up as a small bright white speck in a blue background. In bright sunlight, by standing in the shadow of the deckhouse or a sail, it may be found considerably easier to see Venus in the telescope. Jupiter may sometimes be observed in the daytime, but only with a powerful star telescope.

■ **3. Meridian Altitudes: To Observe The Meridian Of The Sun.** As far as the sun is concerned, we know that it appears to climb continuously from the time of sunrise until noon during the morning hours after which time it appears to dip and continues to fall to until sunset. Noon is the instant that the sun reaches its highest altitude and when it is on the observer's meridian. This highest altitude is called the meridian altitude of the sun, and occurs thus only once daily.

It is customary some little time before the anticipated time of the meridian altitude to get the sextant ready and bring the lower limb of the sun in contact with the horizon. At this time the sun should bear a little to the eastward of true south or true north depending on whether the observer is south or north of the sun respectively. The sun's slow upward progress can now be watched matching it by slight alterations of the tangent or micrometer screw so that the sun's lower limb is kept constantly kissing the horizon. Every few moments we may note the increased reading on the sextant. Eventually the sun appears to stop rising, and there will be no visible change in the altitude for a few moments. When the sun appears to stop rising a little before noon (apparent time) the main point to remember is on no account to reverse the motion of the tangent screw, but to watch the sun intently whilst it remains stationary, until the sun suddenly dips, ie:- when it commences to fall and laps over the line of the sea horizon and the altitude starts to decrease. The instant that the sun appears to dip is accepted as the time of the meridian passage and this time together with the sextant angle should be recorded. It should be noted particularly that as the altitude is required when the sun is perpendicular to the observer's meridian and not to the east or west of it, the meridian altitude must be observed by swinging or rocking the sextant with the pendulum motion described earlier. The meridian altitude may occur before or after noon of ship's time and the approximate time of the meridian passage can be calculated from the estimated longitude.

To Observe The Meridian Altitude Of The Moon, Planets and Stars: As far as the moon, planets and stars are concerned, the meridian altitude should be observed exactly as described above, following the procedure regarding the use of the sextant described earlier in the chapter. The times when these other heavenly bodies will be on the observer's meridian will vary, and must be ascertained in advance from the Nautical Almanac and this type of observation is only valid if there is a clearly defined horizon visible for an accurate altitude.

In every case when a body is on the meridian, it must bear true south or true north, depending on whether the declination of the body shows it to lie to the north or south of the observer's latitude. Therefore, when wanting to find an unseen planet or star to ascertain its meridian altitude, once the approximate altitude has been clamped on the sextant, this should be directed towards the true north or true south points at the horizon. When wishing to observe a second or third magnitude star at twilight for the meridian altitude, it is generally necessary to work out the approximate altitude in advance in this way.

Some of these meridian altitudes using stars and planets will require the observer to work in marginal conditions. It is here that sextant skills really count and these marginal observations could be the critical ones for obtaining a position when making a landfall. As with all sextant work, practice is the key to getting good results. Only experience will let you take a sight quickly or when conditions are marginal. Only experience will allow you to judge the quality of a sight once you have taken it so plenty of practice is the answer.

CHAPTER FOUR

Practical Hints On Taking Sights

The last chapter talked about the need for practice in taking sights. The skill of taking sights cannot be gained from books and it is only actual practice at sea in a wide variety of weather and sea conditions, in daylight and in dark, which will give the experience necessary to make an efficient sextant navigator. Whilst the last chapter covered the main principles of taking sights the following practical hints will help to establish the proficiency necessary to enable the sextant to be used most effectively and to the best advantage.

Preparation For The Sight

The sextant should always be kept in its case with the erect or star telescope fitted and focused ready for use. This may not be possible with some sextants where the box lid cannot be closed with the telescope in situ, but if you can, always have the sextant ready for immediate use. On cloudy days the sun may make only a brief appearance from behind the clouds and if the sextant is ready for use, the efficient navigator should be able to snap a quick sight in a few seconds. Up-dated position information can be vital to safe navigation and it is important not to lose valuable sights through unpreparedness. Whilst waiting for the sun to appear from behind a cloud bank, the sextant should be kept handy in a position where it will not slide about. On a yacht, placing the sextant on a bunk is probably the safest place but the pillow or cushion in the cockpit may be a good substitute. On a larger vessel the chart room settee can be used.

It is necessary to know the height of the observer's eye each time an observation is taken. On a large ship this can vary according to the draught and there can be a considerable difference in the height of eye between the light and loaded conditions on a cargo ship. Most ships will have a table giving the height of eye corresponding to different draughts when taking sights from the bridge, but this height can vary if sights are

Chapter Four

taken from a deck level away from an observer's customary position and a correction in the height will have to be made.

On a yacht the height of eye will not vary significantly through changes in draught but it will vary through the height of the waves in rough seas. The height of eye of the observer can be measured for the smooth water condition and the competent navigator will know the height for different positions such as in the cockpit and at the mast. In rough seas observations will normally be taken when the yacht is on the top of a wave in order to have a clear sight of the horizon. In this situation the height of eye will have to be estimated and the normal way is to add half the estimated height of the wave to the still water height of eye. The height of eye correction is not critical to the accuracy of the resulting position bearing in mind that the best achievable accuracy of the altitude taken in rough conditions is likely to be in the region of 5 minutes. Even a considerable discrepancy in the height of eye correction will not produce an error of more than 1 or 2 minutes.

Guidance about using the telescope and which one to use will vary with day and night conditions. In daylight under all normal conditions when observing the sun, use the erect telescope to get the best results. If the weather is calm and the vessel is steady then use the higher power inverting telescope in order to have a greater magnification and a clearer delineation between the sun and the horizon. When the vessel is moving, this higher magnification will prove impractical as any slight movement could result in the sun or the horizon disappearing from the mirrors.

For morning and evening star observations at dawn and dusk use the erect star telescope under all normal conditions. As with the sun, this telescope gives a wider field of view but less magnification which makes it easier to keep the star in view. The erect star telescope will also produce the maximum brightness on the horizon and of the stars so this should be used for the earliest morning sights and those taken at dusk when the light is fading fast. With experience and in relatively stable conditions the magnification of the inverting telescope can be used when there is still

Practical Hints On Taking Sights

considerable light such as for sights in the very early evening and the latest morning sights. This telescope allows the stars to have maximum brightness.

Taking Sights Generally

The sextant measures the angle between the sea horizon which is sometimes called the visible horizon and the heavenly body. Remember that the altitude of a heavenly body will continue to get larger whilst the object is to the east of the observer's meridian, but will get smaller when to the west of it. Care should be taken to be sure that the horizon is defined clearly. The observer must be very honest and discriminating and to ignore any sights taken when this is not the case. Unless there is a true and unambiguous horizon, the resulting position can be considerably in error. In gloomy weather it is better to take sights from as low a position as possible, as then the horizon is more likely to be clearly defined.

When using the sextant always try to get the observed objects in the centre of the telescope field. This central area is less likely to have optical distortion and of course there is less chance of loosing the heavenly body from sight if it is in the centre.

With a clamping screw found on older vernier sextants always try to keep the slow motion tangent screw near the centre of its run. This will allow maximum adjustment in either direction without having the screw reach the end of its travel. It is very easy to be taking a series of star sights and find the screw travel has been biased in one direction so that the last sight cannot be taken without winding the screw back which can lose critical sights as the light is fading.

It is good practice to always test for index error every time sights are taken. This provides a good check that the sextant is in good order and will give greater confidence in the quality of the sights. Experienced observers generally apply any index error correction to the altitude which is read from the sextant before entering it in the notebook. Whether the correction is applied at this time or later when the sight is being worked out should become a matter of routine and standard

Chapter Four

practice so that there is never any doubt about whether the correction has been applied or not. Of course an even better solution is to have the sextant corrected so that there is no index error, then there is no correction to apply. The other correction to remember is that shown on the sextant certificate. Normally these are minimal and in most cases can be ignored but anything over 1 minute should be applied and in applying this correction, remember that it can vary with altitude unlike the index error.

One of the best ways for a beginner to learn to use the sextant and to build up confidence in its use is to practise taking sights when the vessel is at anchor in a known position. If the calculated position comes within 2 or 3 miles of the known position then this is an adequate result. If there are larger position errors, then the observer should look carefully at his techniques in order to find out where he is making mistakes. Getting an experienced sextant user to take the same observations can be a useful way of finding out where the observing technique is wrong and this can be done during normal sight taking when underway at sea.

Taking sights at anchor will give experience with the sextant under good conditions with the vessel generally stable and the position known. From here the beginner can graduate to taking sights when underway in fine weather near the coast when the position of the vessel can be fixed by bearings of shore objects and the resulting positions compared. One problem with this type of practice is that is may not be possible to get a true horizon as the land may intervene. However the best way to build up confidence in using the sextant is by using practical position comparisons and in good weather with modern instruments, position accuracies in the order of one or two miles should be possible.

In real time navigation situations where the sextant is the primary means of establishing the position when out of sight of the land no opportunity should be lost of obtaining a sight and the resulting position line, particularly when making a landfall. The competent navigator keeps everything in readiness and then when the sun shows signs of appearing through the clouds he is ready to take a quick sight. A practised sextant

Practical Hints On Taking Sights

user will have brought the sun down to the horizon and obtained an altitude within a matter of a few seconds which may often be the only time during which the sun makes its appearance.

When taking sights it is preferable to allow the contact between the body and the horizon to be made by the motion of the objects themselves. In other words, say the sun is rising, the sextant should be set so that the sun appears to be a little below the horizon and careful watch kept until it rises to the horizon and the time taken when it just makes contact. In the case of a falling body watch its fall until it is exactly kissing the horizon.

When taking a meridian altitude of a heavenly body it has to bear either true north or true south as it crosses the meridian or it can be directly overhead in the observer's zenith. When a heavenly body has a low meridian altitude then it will appear to remain stationary on the observer's meridian for several minutes with minimum change in altitude. When the meridian altitude is high the body will only stay at the maximum or meridian altitude for a very short period. When the meridian altitude is nearly 90 degrees, i.e. when the object is nearly overhead, the sun will swoop across the meridian, and appear to touch the horizon all round, so rapidly that it is difficult to decide on an accurate meridian altitude. In this situation, the solution is to observe the altitude facing either true north or south and take the maximum altitude observed. If this turns out to be greater than 90 degrees then the required altitude can be found by subtracting the observed altitude from 180 degrees although any observation taken with the body virtually overhead must be used with a degree of caution.

An ex-meridian altitude may be taken of any heavenly body within the limits specified in the ex-meridian tables, both when the body is rising and falling. The actual observation is taken as already described for each body in the last chapter, and at the same time the exact chronometer time must be noted to the nearest second, so that the hour angle of the body from the meridian may be ascertained. To observe the body after having set the approximate altitude on the sextant, look towards the true north or south horizon, a little to the eastward in the case of a

Chapter Four

rising body, and a little to the westward in the case of a falling body. These ex-meridian altitudes can be a useful source of position lines when the noon sun is obscured and they are often taken before noon as a precaution and only worked out if the noon sun cannot be seen.

If the direct horizon underneath the heavenly body is obscured by land or fog or is of poor quality then it is possible to take a sight using the opposite horizon. The extent of the sextant scale is the limiting factor here and in most cases the elevation of the body will have to be at least 70 degrees for this method to work. In the case of the sun an observation taken in this way has to be worked out with care because if the altitude is subtracted from 180 degrees to get the altitude required you also have to remember that you were in effect taking the altitude of the upper limb of the sun because that was the limb you brought down to the opposite horizon. Another option when the body is high is to take the altitude in both directions. In this case half the difference between the two readings is the zenith distance of the centre but it requires quick sighting if the time difference is not to have a significant influence. This method eliminates dip and index error but if the body is not on the meridian when the sight is taken then the times must be taken of the two observations and averaged to correspond with the zenith distance.

Maximum & Minimum Altitudes

On any course other than east or west, but especially on northerly and southerly courses, the speed of the vessel can affect the observation. When steering towards the sun, the altitude will increase due to the forward motion of the vessel and if when steering away from the sun the altitude decreases. These changes in altitude can be particularly significant on faster vessels, over say 15 knots when the change in latitude with time can be quite rapid. The result is that it may not be possible to get a clear indication of the meridian passage when following the sun's elevation through the sextant in the normal way. The solution is to calculate beforehand the time of the meridian passage and to take the altitude at that time regardless of the movement of the sun in the sextant.

Taking The Time

The time at which an altitude is taken is just as important as the accuracy of the altitude itself. A chronometer is usually carried on board to produce an accurate time reference and this is constantly checked by means of time signals over the radio. Traditional chronometers are beautiful examples of the clock makers art and have to be treated with great care. On board a ship they can get the care they deserve but on a yacht, the often violent motion and the damp atmosphere are not conducive to time keeping accuracy. Modern electronics have come to the rescue in the form of electronic chronometers which are both cheaper and more robust than their traditional counterparts. It is possible to get wrist versions which maintain a high accuracy and for the yacht navigator working on his own they enable sight and time to be taken without moving position.

In order to record the time at which a sight is taken accurately a routine has to be established for the time taking. If two people are available then one takes the sights and the other records the time when the observer shouts 'stop'. Such a system should produce the best accuracy but in many cases it will be down to one person to take the sight and record the time. Some sextants have a stop watch incorporated into the handle of the sextant so that the observer can simply press the stop watch at the time the sight is taken and the stopwatch then counts down until the chronometer time is taken. If the chronometer and the observation position are in close proximity then a simple verbal counting of the time can be accurate enough and practice soon allows the seconds to be counted accurately until the chronometer time can be recorded. On a yacht in rough conditions it may not be practical to make the journey along the deck from the sighting position to the cockpit after each sight in order to record the time and a modern electronic watch is used to take the time at the sighting position and then this watch can be compared with the chronometer to get an accurate time.

Chapter Four

The best solution to taking the time is to have one person for the sight and one for the time. There is less chance of making mistakes and less room for error. It also means that the person recording the time can also record the altitude, allowing the observer to get on with the next sight rather than have to stop and write the altitude down. On a yacht, this is perhaps more critical than on a ship where there can be the luxury of easy movement but for the single-handed sailor the taking of sights needs to be carefully planned and executed if meaningful results are to be obtained. For a two person operation the sextant operator calls out 'stand by' when he has the body close to the horizon which should be acknowledged by his timekeeper. The timekeeper then fixes his eyes on the chronometer face and counts round with the second hand. As the observer makes exact contact and feels he has a good sight then he should shout 'stop'. At this point the timekeeper records the exact second of hearing this call then the minute and finally the hour and acknowledges this to the observer. He should then immediately check the time recorded particularly the minute to make sure that this has not been read in error. When the second hand points to about 50 seconds it is extremely easy to read the next minute by mistake. As one minute of time means an error of 15 miles, accuracy here is important and when concentrating on the second hand it is easy to be more casual about the minute reading. There is less chance of error with a chronometer or watch with a digital readout.

The observer should read his sextant and call out the name of the body observed, its altitude and its approximate bearing in the case of a star or planet. Before moving the index bar of the sextant for the next observation, the observer should again check the altitude to avoid mistakes. The next call should be to "stand by again", when the same operation should be repeated as many times as necessary. The log reading should always be taken at the same time as the sextant observations, so that the dead reckoning position may be calculated. In vessels where the speed and distance run is calculated from the engine revolutions the ship's time should also be recorded. When the final sight is taken the sextant should not be altered. If it is left on the sextant it is there for reference just in case it was wrongly read the first time.

Just as taking sights requires experience so does taking the time. It sounds easy but to get an accurate reading, whether with a dedicated timekeeper or when doing both jobs yourself requires practice and a routine. If you do have an inexperienced person taking the time, then try and glance at the chronometer yourself, before reading the sextant, and note especially whether the minute has been noted down correctly.

When taking sun sights, it is best to observe four altitudes in succession, and take the exact chronometer time in every case. Then take the mean of these four altitudes by adding them together and dividing by four and the mean of the four times and use the mean altitude and the mean time for working out the sight. Should the observer have reason to doubt the accuracy of any one altitude or time or if there is an obvious discrepancy either ignore this observation or plot them all down and draw a curve through them and see if the curve is fair. If in doubt, discard the faulty observation. One good sight cannot be improved upon, but in any but the calmest weather, it is best to take three sights and times and use the mean of the three. The aim should be to get the sights in as short a time as possible, say three sun sights in three to five minutes, and two or three stars in this time.

The Sight Notebook

Accurate results depend to a great extent upon a proper method and routine being used. A notebook called a sight book should be ruled up and every sight taken entered into it, including all necessary particulars. This then becomes a permanent record of all sights taken and provides enough information for sights to be re-worked at a later time if there is some doubt about the calculation or if the sight is particularly important and needs to be checked.

Taking Sights In Rough Weather

Even in a larger vessel taking sights in rough weather may not be easy and this is where skill and experience comes in. In a yacht the problems of using the sextant multiply and the accuracy of positions will inevitably

Chapter Four

decline in rough weather. The main problems are locating yourself safely and securely so the sight can be taken, and the fact that the height of eye is only situated a few feet above the sea, and as the horizon is not far away it is frequently obscured by waves. Care must be taken that the reflected image of a body is brought down to the actual horizon and not to a false horizon or to an intervening wave.

In a rough sea, the actual sight should be taken just when the vessel is on top of a wave when, hopefully the true horizon is visible. Position yourself as high as possible on the vessel although ensure that you don't take undue risks in doing this. Whenever possible, alter course to bring the sun or star either nearly ahead or nearly astern as a vessel usually pitches less than she rolls. When shorthanded or single-handed in a yacht the best solution can be to heave to in order to reduce the vessel motions when taking the sights. In rough weather, 3 or even 5 sights should be taken, and the mean of the altitudes and the times used. Any obviously bad sights should be rejected.

When observing the meridian altitude from a yacht in rough weather, the rise and fall due to the waves causes the sea horizon to fluctuate by the small amount due to the alteration in height of the observer's eye. As it is impossible to keep the sun's limb in constant contact with the horizon, it is preferable to take a number of observations close to noon in quick succession, continuing to do so until it is obvious that the altitude is decreasing. Whilst generally the mean of these observations would be taken as the maximum or meridian altitude, if the mean of the highest and two altitudes on either side be taken as the meridian altitude, the best accuracy will be obtained.

Observations Of The Sun

Sun sights may be taken at any time after the sun has risen about 10 degrees above the horizon. Observations nearer to the horizon than this should not be taken because of refraction which is the bending of the rays of light when they pass through layers of atmosphere of different densities. At other times, refraction must be guarded against most carefully, as heat waves can introduce errors in sights. Refraction through

Practical Hints On Taking Sights

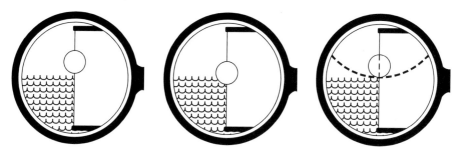

Observing the sun's lower limb. It is first brought down close to the horizon with a coarse adjustment either just above or just below the horizon and then made to "kiss" the horizon.

heat waves is more likely to be experienced in the tropics but it is not easy to establish clear parameters when these heat waves might occur. If a calculated position appears to be in error then bear in mind the possibility that undue refraction has occurred and try to check the position with stellar observations.

When the sun appears from the clouds and shows no apparent sharp outline but is more of a blur then it is possible to take a sight of the centre of the sun. This is only valid if there is an urgent need to get a fix and it means judging where the centre of the sun is. Obviously you do not apply any semi-diameter correction to such an observation. A sight of this type should produce a position which is accurate to within a few miles which will certainly be better than nothing if a fix hasn't been obtained for several days.

In the case of cloud obscuring the sun's lower limb, the upper limb may be used for an altitude. If you apply the semi-diameter correction by subtracting from the altitude you will get the centre of the sun altitude although an upper limb correction table appears in some almanacs. When the sun is at a low altitude, it is most likely to be affected by refraction, so in this case also observe the upper limb which is less likely to be affected by the refraction.

It is important to use the correct shades for the sun and never to use too bright an image of the sun which tends to blur the edges. It is always best to use one shade if possible in order to reduce the chance of optical errors and never more than two should be used. Since we know that when the sun is on the observer's meridian, it must bear true north or south, this provides us with a means of checking the compass error when the altitude is less than about 40 degrees. The compass bearing of the sun is taken at the same time as it reaches the meridian altitude and compared with either true north or true south. The difference between these two bearings is the total compass error for the particular direction of the vessel's head including both variation and deviation.

Observations Of The Stars

When observing stars, beware of using a false horizon. There is usually a more reliable horizon in the evening than the morning. If there is a useable horizon at night and you want to observe stars during the hours of darkness rather than at twilight, it will be found that the planets and bright stars of the first magnitude throw a reflection or halo on the horizon, which makes it extremely difficult to get a clear horizon. It is much better to observe smaller stars of the second or third magnitude, as on a dark but clear night, a smaller star is more easily defined on the horizon.

When taking stellar observations, if a star is not on the meridian at the best time for seeing the horizon, do not wait for the meridian altitude but take the observation at once, noting the exact time by chronometer. You can try and get a meridian altitude later when the time is right, but the earlier sight can be worked as an ex-meridian sight if the meridian sight fails to materialise.

Generally at twilight there will be one star or more on or near the meridian, so try to observe one to the northward and one to the southward. If there are more stars available within the ex-meridian limits try to take sights of them all. The calculations needed to find the latitude from the average ex-meridian tables are relatively easy so that the latitude may be found and checked from several independent stars in a few moments. If the mean of the latitudes is taken as the correct latitude of the vessel, errors will be largely eliminated, and an extremely accurate position line is obtained.

Practical Hints On Taking Sights

When taking stellar observations, it is best to take the meridian altitude observations first as these will take longer and then take the remaining stars as rapidly as accuracy allows. Such observations may therefore, be called "simultaneous" if observed within 2 or 3 minutes of one another provided that the speed of the vessel is relatively slow.

When observing stars in bad or cloudy weather, do not worry if the particular star cannot be identified, but snap it at once (noting the time and altitude in the usual way). The star may be identified later at leisure if the bearing has been taken. The battery operated light fitted to some sextants is a great boon to speed up the reading of the sextant at dawn and dusk and saves having to try and handle a torch. Otherwise the observer must, after having taken the sights, enter the chart room or a cabin which may be brightly lit, in order to read the sextant and it will be some little time before his eyes again become accustomed to the dark.

Latitude by Planet In Daytime

Venus, and sometimes Jupiter, are frequently available during the day and can be used to obtain a latitude. If, as soon as the meridian altitude of the planet has been obtained the sun is observed for a position line, an excellent fix can be obtained rivalled only by a similar fix using the sun and moon.

Because the orbit of Venus is between the earth and the sun, Venus never recedes far from the sun. This means that in most navigable latitudes, Venus is often well placed for obtaining a meridian altitude when either the sun or moon is available to cross with it to fix the position. To observe Venus during the day the sky must be clear blue as even a slight haze will render the planet invisible. Planets sights can be helped by standing in shadow, away from the sun's strong glare. The approximate altitude is calculated and clamped on the sextant, which is directed towards due north or south as appropriate for the meridian altitude.

Observations Of The Pole Star

If the star Polaris, commonly called the Pole Star, were situated exactly at the North Pole, then its true altitude at any place in the northern

Chapter Four

hemisphere would be the latitude. For instance, on the equator the altitude would be 0 degrees and at the North Pole 90 degrees. The Pole Star moves slightly each year however, and revolves around the Pole daily at a distance generally of less than 1 degree from it. Taking a sight of the Pole Star thus becomes a special case of an ex-meridian altitude. The latitude may be obtained from the Pole Star at any time it is visible and the horizon is defined clearly to an observer in the northern hemisphere. In northern latitudes the Pole Star should be one of the choice stars observed at dawn and dusk because it enables a position line to be obtained with the minimum of calculation. Polaris is not a very bright star, being of the second magnitude only, and if the star is visible then the altitude should be observed in the usual way for stars.

When observing Polaris at dawn or dusk however, it may be hard to find. To make things easier the approximate altitude should be clamped on the sextant, which should be directed towards the true north point of the horizon. The altitude may be obtained by working back from the D.R. latitude and a quick, rough and ready supposition is that the

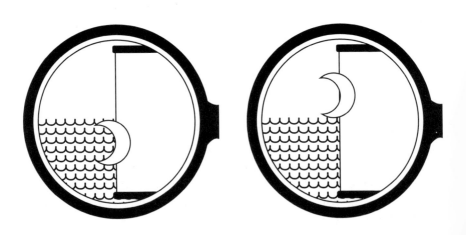

When taking a sight of the moon both the upper and lower limbs can be used, but it is important to make sure that the limb used describes a full arc of a circle particularly when the moon is nearly full.

altitude will equal the latitude. Having found the star in the sextant, it is lined up with the horizon in the usual way. With a single correction table from the almanac the latitude may be calculated.

Observations Of The Moon

Observations of the moon tend to be neglected probably because of the many corrections that have to be applied. However, sights of the moon can provide useful position information and thus have a value for the navigator. The moon may of course be observed during the night, but such sights must be made carefully and treated with caution. In cloudy weather the dark shadows projected on the water below the moon render the actual horizon uncertain. In clear weather the upper edge of the illuminated portion of the sea is the horizon but this tends to be valid only when the altitude of the moon is relatively low.

When observing the moon on the wane, as it disappears from the bottom upwards, the upper limb has to be used. When the moon is near the equator the rate of change in the declination is so rapid that it is unlikely that the actual meridian altitude could be observed with any degree of validity. For much the same reason, and because great exactness would be required, the moon is generally unsuitable to observe for a meridian altitude. During the daytime it may be preferable to take the horizon up to the moon as described already for star sights.

Simultaneous Sights

Simultaneous sights of two or more bodies are of course, not quite simultaneous as a short interval will elapse between observations. Except in very fast vessels if three or four observations are taken within say, five minutes, these so-called simultaneous sights can be assumed to be taken at the same time for plotting purposes although each one has to be calculated with its own respective chronometer time. The prudent navigator will attempt to fix his vessel's position by observation at least three times daily, and the following are the various methods available in the order of preference.

Chapter Four

■ **1. Sun and planet, moon and planet, or sun and moon in the daytime:** During daylight, simultaneous sights of any two bodies are most valuable, and the beginner will find this a good time to commence taking sights, because at this time the horizon is usually clearly defined. Also it may be borne in mind that if the sights fail to work out correctly, they can be repeated which is valuable when learning. A position fixed by crossing position lines from the sun and moon or Venus and the sun are probably the best astronomical fixes available but the hardest to obtain and if achieved they are a feather in the navigator's cap.

■ **2. Simultaneous stars, stars and moon, stars and planet or planet and moon:** The best time to observe stars is undoubtedly at dawn directly the horizon is clear enough. Dawn sights are easier than those taken at dusk as there has been plenty of time available to identify the stars and decide which ones to use. In the evening, stars are frequently not identifiable until the horizon is too dark for accurate sight taking.

Stars and planets need to be carefully selected for observations and where possible, three or four stars (one in each quarter of the compass) are taken in case it is found that the position calculated from just two stars turns out to be some way from the D.R. position. The third and fourth stars can be worked out to provide confirmation of the accuracy of the position. Three or four stars may provide a cocked hat cross and if this is small it can be assumed that the vessel is in the centre of the cocked hat. When taking star sights a frequent source of error is to read the sextant one or five degrees wrong. This shouldn't happen but if there is a major error of this type it can be comforting to have another star sight available to work up to not only produce the correct position but also to identify where the error occurred.

Remember that the ideal set of sights is the one where the position lines cut at right angles. This is manifestly impossible when more than two stars are used, so when taking three or more, try to pick ones with about 45 degrees to 60 degrees difference of bearing between each successive star. Avoid taking sights of stars where the position lines will

Practical Hints On Taking Sights

be parallel or nearly parallel when plotted and which will thus provide a poor cross.

■ **3. 'Running fix' from the sun in the daytime, or the sun and a star at dusk or dawn:** Whenever two sights are taken of the same or different bodies with a run between, it must be borne in mind that the accuracy of the final position depends upon the run between sights and the allowance for distance and current having been made correctly. This method, whilst valid, may be less accurate in a sailing vessel where the speeds can vary which in turn may make the log less accurate.

Single Position Line

When approaching land at the conclusion of a passage or when the land is obscured or when coasting some distance offshore, single sights of the sun or other heavenly body can generate a single position line which can be very useful to the navigator. With a position line, the vessel has to lie somewhere along the line and if the resulting position line lies parallel with the shore it will give the vessel's distance from the shore. If the position line is at right angles to the shore then it will tend to show the point on the shore which is being approached or passed. It is not a hard and fast position but such position lines can be very valuable when combined with soundings and they can help the navigator approach the shore with a greater degree of confidence.

P.V. Sight

A "P.V." sight is one taken when the object is on the prime vertical, that is when it bears due east or west (true). We know that if the object bears due north or south, the position line runs east or west, that is the position line corresponds exactly with a parallel of latitude, and this is the reason why the most accurate latitude is obtained from a meridian altitude. Now when the sun bears due east and west, the position line must run north and south, which means that the position line corresponds exactly with a meridian of longitude. This is the reason why a P.V. sight gives the most accurate longitude. One of the main benefits of a P.V. sight is that

Chapter Four

even if the latitude used in the sight calculations is in error, it should not affect the resulting longitude. In every other case it is necessary to have a correct latitude to work up the position line.

Hints On Obtaining Position Lines

In constantly cloudy weather, no opportunity must be lost of obtaining a sight. Even a single observation will at least give a position line. If the vessel is running north or south the observation should be taken from an object bearing around east or west, as this will give a line of position running in the direction of the vessel's course and indicate whether the vessel is maintaining its course. The same principle applies whatever course the vessel is steering with the main objective being to obtain a position line in the direction of the course line.

This book is about the practical use of the sextant. Whilst this use is inevitably linked directly with the principles and practice of navigation, the navigation aspects of the sextant and the associated calculations have only been touched on when necessary to give a better understanding of the use of the sextant. The reader is referred to navigation books if he wants to continue the study of the calculations for position fixing for which the sextant provides some of the raw material. Here we have looked a little at the principle of position lines and how they affect the selection of suitable bodies for observations but a deeper understanding of position lines is vital to general navigation and to obtaining a good quality fix.

CHAPTER FIVE

How To Read The Sextant

It is very easy to make mistakes in taking the reading of the sextant and this was particularly the case with the older vernier sextants. This source of error has been eliminated to a large extent by the adoption of the micrometer tangent screw but even here it is important to understand the system and how it works if mistakes are to be avoided. Because this is the most common type of sextant these days we will look at how to read this type first.

The Micrometer Tangent Screw

The micrometer tangent screw sextant is so easy to read that it is virtually self-explanatory. It does have a vernier scale, but this is only used for the final stage of accuracy after the degrees and minutes have been read and vernier is used to enable the user to divide the minutes of arc up into tens of seconds which is the general level of sextant accuracy. The degrees are read directly off the arc scale and the large micrometer head divides each degree into sixty which indicates the minutes. It is only at this stage that the small vernier comes into use to divide up the minutes.

For the micrometer tangent screw to work the screw thread in which it operates has to be cut with extreme accuracy. It also has to be maintained at this high accuracy which explains why it has to be kept clean and free from grit and dirt. Each turn of the micrometer wheel corresponds to exactly one degree so that each of the cut threads on the arc corresponds to one degree although in effect it is only half of a degree because of the laws of optics which mean that the reflected ray of light moves through twice the angle which the mirror is rotated.

The arc is graduated to whole degrees in a clear and legible manner. Sextant manufacturers go to considerable trouble to make sextants easy to read to avoid mistakes. The micrometer tangent screw is fitted with a

Chapter Five

large head which is divided into equal 60 parts and reads therefore, to one minute of arc. Although the sea navigator rarely requires to read his sextant to an accuracy of more than to the nearest half minute which can be done by interpolation between the minute marks, a simple vernier is provided against the micrometer head so that the minutes may be sub-divided to 10 second (10") intervals if required. Not all sextants are fitted with this vernier scale to divide up the minutes and opinions are divided about the need for it. Whilst it provides for greater accuracy in the sextant reading, adequate accuracy to the nearest quarter of a degree can be obtained by interpolation which some argue, makes mistakes less likely.

The degrees are read where the index (indicated by an arrow or a mark) cuts the arc. The minutes are read on the large micrometer wheel where the small vernier arrow or a cut mark intercepts with the micrometer wheel scale. When a vernier is fitted the tens of seconds are read where one of the lines on the vernier which is usually to the right of the micrometer coincides with one of the minute lines on the micrometer head.

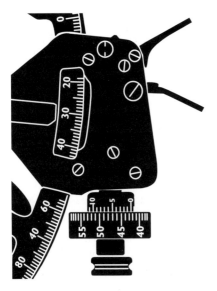

Reading the micrometer sextant. The arrow on the main arc scale shows 29 degrees, the micrometer wheel 0 shows 42 minutes, and the vernier divides the minutes into decimal places, here .5 minutes.

Owing to the many variable factors involved in taking sights such as the vagaries of the horizon, the practical navigator recognises that there is no need to read the sextant quite so accurately, so the custom at sea is generally to read the sextant to the nearest half minute of arc. This makes the whole operation of reading both quicker and less prone to mistakes. With a micrometer sextant the clear scale and marks make it possible to take the reading at arms length during daylight hours. In poor lighting many micrometer sextants are provided with a small electric light mounted on the index arm to give enough light to read the scale. The light is operated by a small push button at the top of the wooden handle. On the finger pressure being released, the light which is powered from a small

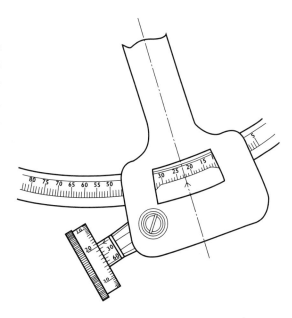

Reading a micrometer sextant where the vernier is divided into seconds rather than decimals.

battery, generally concealed in the handle is extinguished and it is not bright enough to have a significant effect on night vision. The bulb is mounted in a movable arm and by swinging this up the degrees on the arc may be read, and by swinging it down over the micrometer wheel, the minutes and tens of seconds may be read with ease.

The Clamping Screw – Vernier Sextant

As described in the chapter on "How To Take Sights", when the approximate angle is obtained the clamping screw is first screwed up finger tight by the milled head and then exact contact between body and the horizon is made by the tangent screw which is the fine adjustment. The altitude on the sextant may now be read from the main and vernier scales. Because the vernier demands a finely divided scale this is not easy to read clearly with the naked eye even in the daytime. A small magnifying glass is provided, mounted on a swing arm, rather like the light on the micrometer sextant to enable the fine scale to be read. At

Chapter Five

night the same procedure is adopted, but it is necessary to take the sextant to a light so that the scale is illuminated and this is not beneficial for maintaining night vision.

The Endless Tangent Screw Sextant

In this type of sextant, when the approximate angle has been obtained, the index bar becomes automatically clamped to the arc once the finger pressure on the "quick release clamp" is released. As with the micrometer head, exact contact is then made by adjusting the tangent screw. The valuable feature of the endless tangent screw is that, whereas the normal tangent screw can only run to the limit of its thread before it has to be screwed back, thus losing valuable time, the endless tangent screw can be turned constantly without coming to the end of its travel. By turning the tangent screw with the fingers, the index bar may be screwed along the entire length of the arc. It is easy to visualise how this development must have greatly speeded up the taking of sights and for the modern navigator with his micrometer things are even easier and quicker.

The Vernier Sextant

The vernier sextant has largely been consigned to nautical history, but some traditionalists still use them. They have to be admired for their beautiful workmanship and they will produce sights which are just as accurate as those from modern sextants but they require more skill and more time. The vernier sextant works on the principle that the scale on the arc is divided up as finely as possible whilst still allowing the divisions to be read. This normally means that the arc is divided to read to 20 minutes although on some it reads to 10 minutes, making 6 divisions between each degree. The arrow mark on the index arm indicates where the arc should be read from and this is then followed by a series of engraved marks on a scale which runs parallel to the arc.

The purpose of the vernier is to divide up the space between adjacent marks on the arc into much smaller divisions than could be visible by interpolation with the naked eye. The vernier scale will be divided up by matching marks with one significant difference. Whilst the arc scale may have 120 divisions over the same length of arc as the vernier scale, the matching vernier scale will have one less division, in this case 119. Unless the arrow mark which starts the vernier scale is exactly in line with a mark on the arc, it will be one of the other marks and only one which will line up accurately with a mark on the arc. In the case where the arc is divided into 20 minute marks, the vernier scale will be divided and numbered up to 20 with subdivisions in between the whole number up to the 120 quoted above. That mark on the vernier which lines up exactly with the mark on the arc scale indicates the accurate reading and it may be typically 14 minutes and 40 seconds, the vernier scale in this case allowing reading down to 10 seconds or one sixth of a minute because of the scale divisions.

Some cheaper modern sextants still use vernier scales because they are cheaper to produce than a micrometer scale and don't require the very accurately cut worm gear drive. Here they may read only to the nearest minute with the main scale divided to the nearest degree and the vernier scale divided into 60 so that the nearest minute can be identified. A similar type of vernier scale can be found on a micrometer sextant alongside the scale on the micrometer wheel where it is used to give greater accuracy to the reading, usually allowing reading down to 10 seconds. Reading a vernier scale is not easy and understanding the method of reading can be even harder. Once the principle has been grasped that it is the arrow on the index arm which reads the degrees and sometime the proportion of the degrees

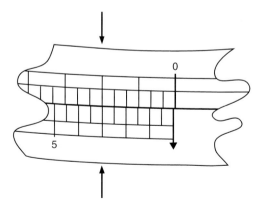

Reading the scale on a vernier sextant. It is the mark on the vernier scale which lines up with a mark on the arc scale which indicates the exact reading. Here it is 00 degrees 3' on the arc.

Chapter Five

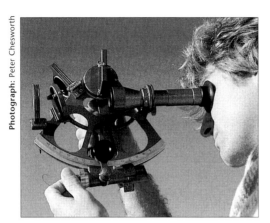

Taking a sight with a micrometer sextant.

directly from the arc and it is the vernier which divides up the spaces between the marks on the arc, then progress is being made and it usually only needs practice to get sufficiently familiar with the system to make quick and accurate readings. When trying to get familiar with a vernier sextant get your reading checked by an experienced navigator and then practice will make perfect. Remember that the degree of accuracy to which it is possible to read off by means of a vernier is always equal to the length of a division on the scale of the instrument divided by the number of divisions on the vernier.

To Read The Sextant

In reading off, whether by day or night, be careful that the magnifying glass is properly focused to the eye, and is immediately over the line of coincidence between the arc and index marks. The light should not fall sideways onto the arc but come from a direction straight down along the index bar to the reading so that there are no parallax problems.

Now to read the sextant, you observe where the index, that is the zero or arrowhead on the vernier cuts the arc. Be careful to make sure you see where the arrow head cuts, because there are usually two or three cuts to the right of zero on the vernier which might be used by mistake. Once this point has been established the reading is taken from the main scale on the arc. Now look along the vernier scale to identify the point at which one of the marks on the vernier scale is exactly in line with a mark on the arc scale. This is where you take the reading from the vernier scale and if it is divided into 20 numbered divisions then this will represent 20 minutes and from this you can work out just where the lined up reading comes on the scale. It should be noted that about 3 cuts on the vernier will appear almost to coincide with the arc cuts and two observers will not always read the sextant exactly to the same 10 seconds; whereas all observers would agree to the nearest half minute.

With a micrometer sextant everything is so much easier with the main divisions on the arc which is usually divided into degrees, being simply read off the arc scale where it lines up with the arrow. Then you read off the minutes directly from the micrometer scale and it is usually possible to interpolate on this scale to a quarter of a minute or even 20 seconds. This is generally more than adequate accuracy but the vernier scale added inside the micrometer scale can make this interpolation more accurate.

Normally when we talk about reading the sextant we are talking about reading it on the arc, ie:– reading it when the reading point is above zero. However there are times when the reading is off the arc, such as when ascertaining the index error or taking a vertical angle of a shore object. On this part of the arc the sextant is reading 'backwards' and it needs explaining how to read the sextant on this part of the arc.

With a vernier sextant the arc reading is taken in the same way as described above, using the zero or arrowhead to read from the arc scale. For the more accurate reading provided by the vernier, the vernier scale has to be reversed in your mind so that the zero of the vernier scale is at the other end from the marked zero, normally marked 20. Now as we are reading backwards from the right the vernier should also be read backwards, increasing from the right. An alternative way to read the vernier off the arc is to read it in the usual way and then subtract the vernier reading from 10 or 20 depending on the scale divisions. This requires more mental agility and increases the chance of mistakes. The best thing is to use the method you understand and stick to it so there is less chance of errors. Many users prefer the second method because the vernier is then always read the same way from the same end which at least introduces some consistency.

With a micrometer sextant, the degrees are read off the arc scale in the normal way but the micrometer works in reverse when it is off

Chapter Five

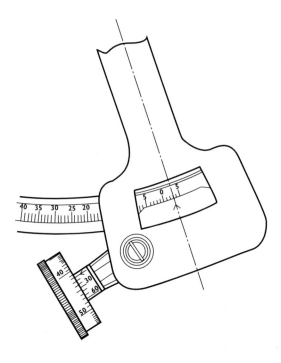

Reading the micrometer sextant off the arc. Here the reading is 3 degrees 17′ 50″ off the arc.

the arc. You read the micrometer and then subtract the reading from 60 to get the minutes and portions of minutes. If you want to get ultimate accuracy by using the micrometer vernier then the vernier has to be read in reverse as above. It can get complicated and it is best to interpolate rather than use the vernier scale when working off the arc. It is hoped that reading the sextant has become clear from the above but like most things associated with the sextant and its use it is only practice that will give the confidence necessary to be able to read the sextant quickly and accurately.

Decimal Graduations

Many modern micrometer sextants are graduated to read to decimals of a minute rather than to tens of seconds. This means that the vernier scale associated with the micrometer will read to decimal places, usually .2, .4, .6, .8 and the lower marking as 1'. This trend is in accordance with the way in which information is now presented in nautical almanacs. The difference is not very great but if you come new to a sextant then it pays to establish which way the sextant presents the seconds information. The same thing is happening with charts and with the electronic positioning fixing instruments so it is logical for the modern sextant to follow suit.

CHAPTER SIX

Errors & Adjustments

Whilst a sextant is built as a precision piece of measuring equipment the designers build in a range of adjustments which allow the owner to fine tune the accuracy and get the best possible results. Although the navigator should take good care of his sextant, it is often not realised that its very construction makes it quite a delicate instrument. This delicate nature of the sextant is more to do with the very accurate angles it is required to measure rather than the actual construction which is quite robust. Because of the high level of accuracy only a very small change in the sextant parameters can affect the overall accuracy and the adjustments built in to the sextant are aimed at maintaining the accuracy when small changes occur to the relationship of the mirrors and the measuring arc. A competent navigator will keep his instrument in the best working order and the highest state of adjustment to reduce the impact of errors.

The errors of the sextant are all concerned with optical parts of the instrument. All the parts of the sextant should be well related to each other, the index glass, horizon glass, and all shades should be perfectly flat and smooth, whilst the sides of these glasses should be parallel to each other. The arc should be perfectly graduated and the index glass perfectly centred.

These errors of the sextant will be more readily understandable if they are divided into two parts which may be termed non-adjustable errors, those which the navigator is unable to correct for himself, and adjustable errors which he can check and should correct at regular intervals. Correction is encouraged by the provision of the adjusting pin or key in the sextant box.

Chapter Six

Non-Adjustable Errors

These errors, which are all really errors of construction, are those which the navigator has to accept and can do nothing about except apply corrections to his calculations:

- **1. Graduation Error:** The arc graduations of the sextant may be inaccurate through faulty cutting. This is an error which cannot be discovered by the navigator as any faults are likely to be extremely small and would require high precision and specialised equipment to detect them. This is a fault in construction that cannot be remedied and the navigator has to rely on the sextant certificate from the maker that any graduation errors are minimal and that the corrections shown on the certificate incorporate them.

- **2. Centring Error:** This is another error of manufacture, and again there is no means of ascertaining whether it exists or for adjusting if it does. It is caused by the centre of the arc not coinciding with the centre of rotation of the index bar, in other words, when the index bar does not pivot at the centre of the arc. Such an error would vary at different positions of the index bar, generally increasing with the angle measured, but it tends to be a very small error if it exists at all, and has no appreciable effect on small angles. Like the graduation error, any centring error should have been detected during the initial testing and will be tabulated on the maker's certificate.

- **3. Shade Error:** This error is caused if the two faces of the shade glass are not ground exactly parallel. It cannot be adjusted but if it is suspected it may be ascertained by taking an observation of fixed objects (not heavenly bodies) with the particular shade, and noting the reading, and then taking an observation without the shade, and comparing the reading. Any error found should be carefully recorded and if a shade is found to be defective in this way it should either be replaced or not used. Whilst a correction could be applied, there

Errors & Adjustments

must remain doubt about whether the shade error exists over the whole surface of the shade and whether the error found is valid at all times when using the shade.

Another way of checking for shade error on both the index and horizon shades is by making contact with the direct and reflected images of the sun, using the inverting telescope and the darker telescope eye piece, not forgetting to adjust the rising piece to make the two images equally bright. If, on removing the dark eye piece and turning in the shade to be tested, and at the same time fitting the lighter eye piece if necessary, the sun's images remain in contact, the shade is accurate. If however, the images of the sun do not remain in contact then the difference in the sextant readings of their being in contact with and without the shade is the shade error for that particular shade or combination of shades.

The question of whether the error in each case is positive or negative must be carefully considered. If the index bar has to be moved higher on the arc proper to obtain contact with the faulty shade, then this error has to be subtracted and it is a minus error. If however, the angle to obtain contact with the faulty shade is lower then the error has to be added and it is a plus error.

Shade error can of course be caused by the shades becoming loose or damaged by twisting, but such a condition should be apparent at once to the careful observer. In this case the use of the shades should be discontinued until they may be properly tested and repaired, and one or other of the dark eye pieces fitted to the telescope and used. It is important to do this, as by using the telescope eye piece any shade error is eliminated. Do not forget to move the up and down screw, the rising piece, until the images are equally bright.

It should be pointed out that in modern sextants shade error is rare, and it can be ignored in any sextant with a certificate from a reputable manufacturer. The possibility of shade error emphasises the need to obtain the best quality sextant that can be afforded. All of these non-adjustable errors are more likely to be found in cheaper sextants where the manufacturing standards may be slightly

Chapter Six

lower but even on lower cost sextants, the amount of any errors of this type is likely to be very small.

The Sextant Certificate

In checking a sextant prior to issuing its certificate it is mainly these non-adjustable errors which are checked. The certificate may state that:–

The divisions on the arc has been examined at a number of points along the arc and found free from material error. The following corrections, in addition to the index correction, should be applied to the readings of the arc. Then follows a list of the corrections to be applied, either plus or minus. These corrections are generally quite small, rarely more than 20 minutes and unless high accuracy is sought they can generally be ignored for practical purposes. The sextant certificate on which they are recorded is usually attached inside the lid of the sextant box.

Adjustable Errors

There are four errors of the sextant which the navigator can control and which he can make adjustments for:–

■ **1. Error of Perpendicularity:** The index glass must be perpendicular to the plane of the instrument. If this mirror is not square or perpendicular to the plane of the instrument it is easy to see how it will not reflect the ray of light accurately to the horizon mirror and then to the telescope. To check for error of perpendicularity place the index bar to about half way along the arc at approximately the 60 degrees mark and hold the sextant horizontally face up with the index glass close to the observer's eye. This puts the arc of the instrument away from the observer and removes the telescope so that there is a clear view of the arc both actual and reflected through the index mirror.

Now direct the eye obliquely into the lower portion of the index glass and observe whether the reflected arc as seen in the mirror to the left is in the same plane as the arc seen by direct vision close outside the mirror to the right. The actual and reflected arcs should form one

Errors & Adjustments

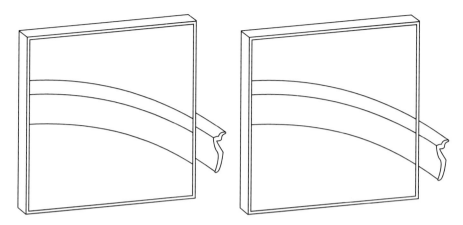

Above left: *Viewing the actual and the reflected line of the arc to check whether the index mirror is at right angles to the plane of the sextant. Here an error exists.*

Above right: *Both actual and reflected arcs form a continuous line to indicate that there is no error of perpendicularity.*

continuous line if no error exists and the index glass is perpendicular, and therefore reflecting correctly. If however, the arc appears broken where the images meet, that is the line of the reflected arc is seen in the mirror higher or lower than the actual arc, then there is an error of perpendicularity and the ray of light from the observed body is not being reflected to the horizon glass parallel to the plane of the instrument. To try and rectify this the observer would tilt his sextant and thus the angle obtained would not be properly perpendicular to the horizon.

If the reflected arc seems to drop below the actual arc then the index glass leans backward at its top, but if it is seen to rise above the actual arc then the index glass is leaning forward. To rectify the error the top adjusting screw at the back of the index glass must be tightened or loosened. This screw presses against the back of the mirror to change its angle and it is adjusted until the true and reflected arcs form one continuous line to indicate that the mirror is upright.

Before making the adjustment any cap covering the index glass adjusting screw should be removed to expose the screw itself. There are three types of adjusting screw commonly used on the sextant. One type is an ordinary screw turned by a small screwdriver but the more usual

Chapter Six

type has a head with several holes drilled through it, into which the small ball headed adjusting pin is inserted. The third type has a square or hexagonal spindle which is turned by means of a special key supplied with the sextant. The second and third types are preferable to the first because the key or pin stays in place and can be turned without taking your eye off the arcs making the adjustment a smooth operation. In making the adjustment hold the sextant in the horizontal position with the index mirror towards you. Then by gently turning the adjusting tool one way or the other the true and reflected arcs may be brought into one continuous straight line.

■ **2. Side Error:** The horizon glass must also be perpendicular to the plane of the instrument and this second adjustment which is known as side error is to adjust this mirror. The error produced if the horizon mirror is not perpendicular is similar to that of the index mirror and it means the ray of light is not being reflected accurately. To check whether any error exists, the sun, a star or the horizon can be used.

The best way in which to carry out this second adjustment is by using a fairly low lying star that is not too bright. First clamp the arc at about the zero mark, fit the inverting telescope, and holding the sextant vertically in the usual way for observations look directly at the star. Turn the tangent screw or micrometer wheel so that the index arm moves across the zero of the arc, ie:–

alternatively a little on and off the arc. The reflected image of the star seen through the silvered section should exactly pass over the direct image of the star. If the true and reflected images of the star coincide exactly as they appear to pass one another then no adjustment is required. If however, they appear to pass to the side of one another and do not coincide, then there is side error and correction is necessary.

To make the correction, set the index bar exactly at zero using the micrometer wheel or vernier to ensure that the zero is accurate. The mirror is then adjusted by using the centre screw at the back of the horizon mirror, checking after each adjustment whether the true and

Errors & Adjustments

reflected stars have been brought into coincidence. This can be a bit of a fiddly job as it is not always easy to make this adjustment whilst viewing the star continuously but once the required direction of adjustment is established it is a quick matter to make the final alignment.

As there are two adjusting screws at the back of the horizon glass rather than the one on the index glass it is important to understand what is happening. With this second adjustment the horizon glass is being made perpendicular to the sextant frame and therefore the glass is being pushed forward or back a little at the top as the screw is adjusted. To achieve this it is obviously the screw in the vertical centre line which may be either at the top or the bottom of the glass which is used.

During the daytime the sun or the moon may be used as the adjustment reference but they are not as satisfactory as using a star. When using the sun fit the dark eye-piece to the inverting telescope. This avoids using the shades and eliminates any shade error which might exist.

Now hold the instrument vertically and look directly at the sun with the index at zero. Now move the tangent or micrometer screw generously backwards and forwards, and if the image of the sun passes directly over the actual sun there is no side error. If however, it passes either side there is side error and it has to be eliminated as described above. The same procedure is adopted with the moon but its uneven shape can make it less reliable as a check.

The sea horizon can also be used as a reference but it will not give the same accuracy as the star method. For this set the sextant exactly to zero on the arc using the micrometer or tangent screw for accuracy, insert a telescope and holding the sextant horizontally, look at the sea horizon. Now both the reflected image of the horizon and the true horizon on either side of the mirror should be in line if there is no error. If the horizons are not in line then side error exists and adjustment should be carried out as described above to bring them in line. The reason why this method of adjustment is not recommended is because the sextant has to be held horizontally which (except in the case of surveying sextants or sextants used chiefly to observe horizontal angles) is not the position in which it will be normally used.

Chapter Six

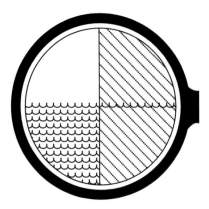

Lining up the true and reflected horizons in the horizon glass to check for index error.

■ **3. Index Error:** The horizon glass must be parallel to the index glass when the index is at zero. It is easy to see the logic of this requirement because it is by measuring the difference in angles that the required altitude is obtained. If they are not parallel at zero they will not be accurate over the rest of the range.

The error can be checked either by using a star, the sun or the sea horizon. When using a star, clamp the index arm at exactly zero, again using the micrometer or the vernier to verify the precise zero. Observe a star with a low altitude using the high power inverting telescope and holding the sextant vertically in the usual way.

If there is no error the coincidence between the true and reflected stars should be perfect both vertically and horizontally. It is the same when the horizon is used. With the sun it is not easy to tell if they are in line and the best way is to make them touch one above the other and then reverse the positions. The same readings should be obtained on and off the arc and if there is a difference then index error exists. If the reflected image of the star or the horizon is above or below the true image, ie:– if a double star is seen, then index error exists so the true and reflected stars must be brought into coincidence by using the second of the adjustment screws at the back of the horizon mirror which is the one at the side rather than the top or bottom.

It will be easy to decide which of the two adjustment screws on the back of the horizon mirror should be used if we consider that we are trying to make the two glasses parallel when the index is at zero. This means that the edge of the glass must be moved round to achieve this which is the opposite to what was done in the second adjustment. Therefore the screw to use is the one at the side of the horizon glass. The

Errors & Adjustments

sea horizon may also be used for this adjustment exactly as described for the second adjustment, but by holding the sextant vertically.

It must be remembered that the second and third adjustment screws work on the same mirror and, to a certain extent the two motions induced by the adjusting screws will be dependent on each other. Having made the third adjustment for index error it is a good plan to go back and check the second adjustment in case it has been affected when the third adjustment was made. It can be a tricky business getting both of these adjustments made in order to eliminate both of the errors completely and the adjusting screws on the horizon mirror may have to be worked on alternately and very gently before both adjustments are completed satisfactorily. Even then it may not be possible to eliminate both errors completely.

If this is the case the best solution is to halve the errors. After having set the index at zero, take out half the side error by the top or bottom screw. This will probably alter the index error, so take out half the index error with the side screw, then halve the remaining side error and continue this until both have been eliminated. It is important not to neglect to do this as otherwise side error may be left in the glass. For the above reasons then, unless the third adjustment is considerable, say more than 3 minutes it may be advisable not to correct for it at all, but to ascertain its amount and allow for this at each observation. It is one more correction to apply and obviously the more corrections you have to make in your calculations the more chance there is of errors.

Owing to the danger of the instrument being unwittingly put out of adjustment by the adjustment screws working loose during handling of the sextant the adjusting screws should be worked as little as possible. However the sextant should be checked at regular intervals for errors for this very reason and once you become familiar with the checking routine it only takes a very short time to make the checks.

■ **4. Collimation Error:** The axis of the telescope and hence the line of sight through it must be parallel to the plane of the instrument. The line of collimation in an inverting telescope is the imaginary straight line

Chapter Six

which passes through the centre of the telescope, the object glass and the eye glass and intersects at right angles with the cross wires placed at the focus of the eye-piece. It is the optical axis of the telescope, and must be parallel to the plane of the instrument.

The elimination of the collimation error is the one that has presented most difficulties to the navigator owing to the somewhat complicated way of testing for the error and adjusting it. This error is due to the fact that the reflected ray from the silvered part of the horizon glass is not received in the same vertical plane as that from the horizon and it can cause the observed angles to be too high.

Fortunately however, this error is rarely seen, and in modern well made sextants it is almost impossible for the telescope to get out of collimation unless of course the instrument has been knocked and the rising piece bent, which would be obvious to a careful user. If the axis of the telescope is not parallel to the plane of the instrument, then if this condition was so small as to be unnoticeable, it would be a small error indeed, probably under a minute.

Before adjusting for collimation error the first three errors must be eliminated. To make adjustment, the two cross wires of the inverting telescope must be placed parallel to the plane of the instrument. This is done by observing a low lying star and placing the index arm 3 or 4 degrees on either side of zero, thereby separating the two images of the star, so one is seen near the top of the field of view and one near the bottom. Bring the edge of the vertical wires into contact with the top image and by revolving the eye-piece in the telescope, the two images of the star should be made to appear in contact with the same edge of the wires. The cross wires should now be parallel to the plane of the instrument, and the tube and draw of the telescope may be marked for future reference if desired.

Now with the cross wires in the position described above, select two heavenly bodies, preferably two stars about 90 degrees-120 degrees apart, and bring the reflected image of one and the true image of the other into coincidence on the wire nearest the instrument, the lower wire. Clamp the index and by slightly moving the position of the sextant

Errors & Adjustments

make the coincident stars appear on the wire farthest away from the instrument (the upper wire). If the stars still remain in contact, no error exists, but if not collimation error does exist. It requires skill in sextant handling to bring the stars together in this way and to make the check. The sun and the moon when at a considerable distance from one another are excellent heavenly bodies to use on occasion for the adjustment of this error. Bring the darkened image of the sun to touch the moon, which is viewed directly at the middle point of the lower wire. Then by moving the sextant, bring the point of contact to the middle point of the upper wire, where there should be exact contact also. If the images appear separated at the upper wire, the eye piece end of the telescope rises up, if they overlap it droops down. In the first case, ease the top screw and tighten the lower screw, and do the opposite if the two objects appear to overlap at the top wire.

When using stars if the two images appear to separate when transferred from one wire to another, then the object glass (the one furthest from the eye) is too near the frame. The adjustment should be made by the two opposing screws in the double collar of the sextant which hold the collar against the exterior ring. By slackening the screw farthest from the frame of the sextant ie:– the top screw, and tightening the other one, the bottom screw, the telescope can be brought into line. If the two images overlap this procedure should be reversed. Whilst the above methods can show collimation errors by observation of heavenly bodies it is possible to make a visual check simply by looking at the instrument. Screw the inverting telescope in place, and hold the sextant carefully in such a position that an inspection will show clearly if the telescope is parallel with the plane of the instrument. An error as small as 1 degree may be seen easily by the eye in this way and will serve as a routine check. Whilst there are several other methods of testing for the collimation error, the above methods should be quite sufficient. Some modern sextants have no means of carrying out an adjustment for this error anyway as many manufacturers regard it as unnecessary. The above tests cannot be made with the star or prismatic telescopes, which are not fitted with cross wires. This combined with

Chapter Six

the absence of adjustment screws means that most modern navigators ignore this error.

To Find The Index Error

Although the name is used to describe the third error of the sextant the term index error is also used to describe the amount of error left on the sextant after the four errors described above have been adjusted and as far as possible, eliminated. As mentioned under the third adjustment, if the error is found to be considerable, say more than 3' then the adjustment is carried out; but if not, the screw is left untouched and the index error ascertained and allowed for. It is important always to know the index error of a sextant, and whilst the navigator would only check over the adjustments occasionally, he should constantly check his index error. Although strictly speaking, this should be done at each observation, the observer should certainly check it every two or three days at sea.

It should be understood that, whenever a sextant has an index error, this must be added to, or subtracted from, every altitude or angle taken with the instrument, regardless of the body observed or the altitude obtained. If this error is not removed by adjustment it will affect all observed angles alike. There are three ways of finding the index error, namely:–

The sextant and its box. The accessories are stowed in the front of the box and its sextant certificate is in the lid.

Photograph: Peter Chesworth

Errors & Adjustments

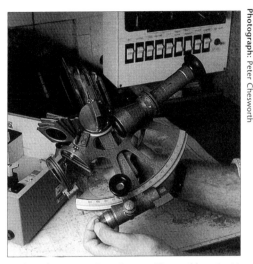

Photograph: Peter Chesworth

Reading The sextant. Note the magnifier on this sextant to ensure a clear reading of the one degree arc divisions.

- By observing a star.
- By observing the sea horizon.
- By observing the sun's diameter "on" and "off" the arc.

The first and best way is to observe a star with a low altitude; the second and usual way is to use the sea horizon, and the third way is by the sun. If the weather conditions are such that all three methods may be employed, the same result should be obtained in every case which can act as a good check. The best results will be found by using a star, because as a star has no appreciable diameter no correction for semi diameter is necessary, and no shades need be used, therefore shade error should not arise.

When using a star set the index arm at approximately zero (a few minutes one way or the other off zero) and point the inverting telescope at a low lying star. Make the true and reflected stars exactly coincide and read the vernier or micrometer head. If the reading is exactly zero, of course there is no index error. Should the reading be "on" the arc, eg 2'20", then this would be 2'20" to be subtracted from the taken altitude. Should the reading however be 2'20" "off" the arc, then this would be 2'20" to be added to the observed angles.

If the side error has not been entirely eliminated, then the stars will appear at an oblique angle to one another. An endeavour should be made to eliminate side error, but if this is not possible the reflected star should be brought down dead alongside the direct star, and any angle on the sextant would then be the index error.

Chapter Six

When using the sea horizon to find the index error set the index arm exactly at zero and then using the inverting telescope and holding the sextant vertically look direct at the sea horizon. Should the true and reflected horizons appear to form one continuous line then no index error exists. Should they not be in line, then by moving the vernier or micrometer wheel, make an exact continuation of the reflected with the true horizon. Read the sextant and ascertain the index error which must be allowed for exactly as described above. To ensure accurate results care must be taken to see that the sea horizon is clearly defined. Finding index error by the sea horizon should be used only as a quick check or when heavenly bodies are not available.

When using the sun the index error may be found both by vertical and horizontal angles in which the sun's diameter can be measured both "on" and "off" the arc. It will be found generally more convenient to observe the horizontal diameter because it is not so easy to observe the sun at high altitudes, and at low altitudes the vertical diameter may be distorted by refraction.

Now if the index be placed at 0 degrees when there is no index error, the centres of the reflected and direct images of the sun will be coincident; and when there is an index error the centres will not coincide.

If we could exactly superimpose one image upon the other, the reading would give the divergence of the sun's centres or the index error of the sextant. This cannot be done correctly however, so the following method is adopted. Using the inverting telescope, and the necessary shades, set the index about 32' on the arc which is approximately the diameter of the sun. This is much quicker than clamping at zero, and moving the vernier along to 32' by the tangent screw which could come to the end of its thread. Let the left limb of the reflected image of the sun be brought into exact contact with the right limb of the direct image.

Read the angle of the sextant and note it down. Now bring the reflected sun across the actual sun until the opposite limbs are brought into contact, that is, the right limb of the reflected image into exact contact with the left limb of the direct image which will be at about 32' off the arc). Read and note down the angle once more. If, as is generally the case, one of these readings is on the arc proper and one off the arc, half their difference is the index error. If the greater of these two angles is on the arc, then the index error must come off and is subtracted from the altitude but if the greater of the two angles is off the arc, then the index error must be added.

If the sum of the two readings is taken and divided by four, this should agree with the sun's diameter for the day, as given in the Nautical Almanac, and be a good check on whether the readings have been taken correctly.

Whilst the index error can be found from the sun it is a more complex business and it is preferable however to use a star for ascertaining the index error. If none of the above methods are available a distant mark could be used .

In summing up sextant errors it is generally safe to say that all of the non adjustable errors may be ignored in any modern sextant of good make except those shown on the certificate. However these are generally too small to worry about for practical purposes. Adjustments need to be made for the remaining errors but if this is not practical then find the index error and apply this to the altitude observed at the time. There remains the possibility of a persistent error in the sextant caused through a knock or other damage. This should show up when the index error is checked but if there is concern about the accuracy then have the instrument checked out by the makers or a repairer.

CHAPTER SEVEN

Sextant Telescopes & Other Accessories

In the past good quality sextants would come supplied with a number of telescopes, each one designed to allow the best observations to be made in the easiest manner under specific conditions. The more expensive the instrument, the more superior would be the telescopes provided in order to allow observations in marginal conditions. Now with improved telescope technology most modern sextants come with just one telescope. Alternative telescopes are available with many modern sextants as an option but these are generally only required for specific observation tasks. Here we will look at both the old and the modern approach to sextant telescopes so that all types of sextant currently in use are covered.

No attempt will be made to go into the theory of the construction of the different kinds of telescopes, but we will look more at the various practical points that arise for the navigator who wants to get the best out of his sextant and to be able to optimise the viewing of both heavenly bodies and shore objects.

In any telescope the end farthest from the observer, which is generally the larger end, is called the object glass, and the end towards the observer, the smaller end is called the eye piece. With the older type of sextant telescope the lenses in this eye piece are fixed in a separate tube which is generally called the "draw" of the telescope which slides in an outer tube. This movement of the eye piece in or out allows the telescope to be focused for different eyes. On modern telescopes this focusing arrangement is generally by means of a screw adjustment on the eye piece similar to that found on binoculars. The screw adjustment allows for finer focusing and is also less likely to be unwittingly adjusted by eye pressure. With the draw tube system, when the telescope has been focused and the objects have been made to appear well defined and distinct to an individual observer, it is often the practice to make a circular pencil ring round the "draw" at the correct position for his eye. Then when experience has shown that this mark represents the best position for this particular observer's eye, the pencil mark can be permanently cut in with a sharp knife, so that this may be felt with the finger nail to ensure correct focusing of the telescope when taking star sights in the dark. It simplifies daylight observations also, as it may be seen easily and avoids the necessity of the telescope requiring to be focused each time. All the draws of the various telescopes should be marked in a similar manner. With the screw type of adjustment, there is usually a scale against which a mark moves and the focusing point can be highlighted on this scale. On these modern telescopes with the screw adjustment the focusing is usually not critical as the sextant can be stowed in its box without altering the telescope whilst with the draw tube type, the draw often has to be pushed in to make it more compact for stowage.

Blank Tube Or Sight Vane

This blank tube is simply a telescope with no optics and serves the purpose of a sight vane which tends to be only used when taking shore angles to keep the line of sight from the eye, parallel to the plane of the instrument. It is not usually found on modern instruments where the general purpose telescope is equally suitable for shore angles.

Telescopes

There are really three principal types of telescopes used with sextants – the Galilean (which shows objects the correct way up), the inverting telescope (which shows objects upside down), and the prismatic.

The image seen by a Galilean telescope is erect and this telescope is known to navigators as the erect telescope. It is also frequently called the star telescope. Such a telescope cannot be fitted with cross wires.

The image seen by an astronomical telescope is upside down or inverted, and this type of telescope is known as the inverting telescope. Because it can be fitted with cross wires, it may also be called a collimation telescope.

Chapter Seven

These two distinct types of telescope are supplied with most older types of sextant and they cover all usual requirements, which are for a low power telescope with a wide field and good light transmission, and for a high power telescope for special purposes. On modern sextants the standard telescope is generally one which combines relatively high power or magnification with a reasonably wide field of view so that one telescope can do both jobs.

The magnification or power of a telescope is generally written as, for example, 2 x : this (the multiplication sign) meaning that any object is magnified twice the size. As a general rule the higher the magnification the smaller the field of view, hence the reason for having two telescopes on older sextants. So as to retain a fair sized field of view, the magnification of telescopes is kept low and seldom exceeds 5 x (5 times).

The Star Telescope

The lower power erect telescope (or star telescope) is the short conical bell-shaped telescope and is used for all general observations. In older sextants of course, the erect telescope was purely a straight one and was employed for sun sights almost entirely. Since it was recognised that navigation by stars was in every way more desirable than using the sun exclusively, this low power telescope has been improved to make it of the best possible utility for star work. Consequently it is generally now called the star telescope. Its power is usually 2 x, 3 x or 4 x with most modern sextants using a magnification of four for this telescope.

The main improvement which has been incorporated into modern instruments has been in making the object glass larger. With its magnification of four these modern telescopes will have an object glass of about 40 mm diameter. The purpose of the large object glass is to overcome the restriction of the field of view which results from the erect eye piece and to give increased light transmission. The diameter of the object glass is a measure of its light collecting qualities and a modern sextant telescope of this type will be described as a 4 x 40. It should be pointed out that a high powered telescope is of no special value at night, because a fixed star's image cannot be magnified. A star will always appear to be a mere point of light, so for this reason the magnifying power of the star telescope is not high. The use of a large object glass telescope however, is very helpful in creating both the brightest possible star and horizon. The larger field of view can also help in finding the star. However, care should be exercised when using such a telescope in bright sunlight. Eye damage could result from looking at the sun directly through such a telescope and shades must always be in place before looking at the sun. The star telescope should also be employed when observing angles between two terrestrial objects unless a special telescope is provided.

The High Power Inverting Telescope

The inverting telescope will often have two draw or eye pieces, each with a different power from the other. Each of these eye pieces is fitted with cross wires at its focus to define the line of collimation, which is the line joining the focus to the centre of the object glass. The high power eye piece generally has two cross wires and the low power eye piece has four. The high power telescope contains fewer lenses than the erect telescope and since it is only used for observations of the heavenly bodies (or the horizon) the fact that it shows things upside down or inverted does not matter. It is especially useful for making the adjustments of the sextant, using the higher power eye piece to give finer adjustment, and it would also be employed if an artificial horizon was being used on shore. The higher power eye piece used for this purpose would have a 9 x or 10 x magnification.

The lower power eye piece (generally 6 x magnification) has a wider field of view and gives greater light transmission so this would be employed for all observations of the sun where the deck is steady enough for the observer to use it. It gives a much larger image of the sun than the star telescope and therefore should always be used when conditions are favourable for taking the sun's meridian altitude because the sun's limb will be much more clearly defined. With a steady deck an experienced navigator will use this inverting telescope for ordinary observations of the sun when conditions are favourable, the high power eye piece being

Chapter Seven

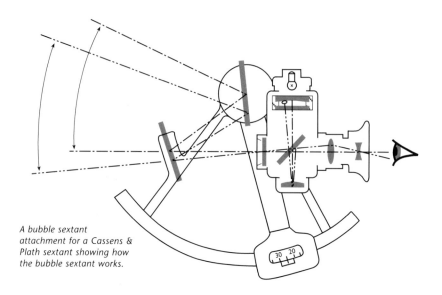

A bubble sextant attachment for a Cassens & Plath sextant showing how the bubble sextant works.

used when the horizon is bright and the ship steady. On small craft this high power telescope will not have a lot of relevance because the vessel is unlikely to be steady enough to use it effectively at any time.

The Prismatic Telescope

The prismatic telescope gives the same benefits which are found with the inverting telescope, but it gives an erect image. For this reason therefore, it is of special use when taking horizontal angles of terrestrial (or shore) objects. It should also be used with the bubble artificial horizon. Its power is usually 6 x and its object glass or field of view is usually 30mm. It is, in effect one side of a pair of low powered binoculars. Because it gives an erect image like the star telescope, it is easy for the observer to change from one erect telescope to another.

Alternative Telescopes

Binocular glasses (having two eye pieces like ordinary binoculars) are fitted to some sextants, especially for taking observations of stars. They do not appear to be popular at sea however, sea navigators preferring to use the familiar star telescope. If binoculars are used however, care must be taken to fit them correctly to the collar, or errors of collimation may occur.

An alternative type of telescope has a compass built in to the unit so that the observer can view the compass and the heavenly body at the same time. The value of the compass comes when the bearing and height of the star have been pre-calculated as can be the case at dusk star sights. By pre-setting the calculated height on the arc and pointing the sextant towards the correct bearing shown on the compass, the star should appear in the telescope.

Some manufacturers offer a bubble sextant telescope. The bubble sextant creates an artificial horizon which allows sights of stars to be taken when the real horizon is not visible. Bubble sextants generally require a stable platform and are not usually suitable for use at sea, but the units incorporating the artificial horizon or bubble into the sextant telescope can be used at sea in good conditions. Built in illumination enables these units to be used at night which opens up the possibility of making observations on demand as long as there is a clear sky. The bubble sextant telescope is an expensive accessory and its use can rarely be justified except for specialised requirements.

Sextant Acessories

There are one or two accessories that may be used with the sextant, and when purchasing a new sextant, consideration should be given to the fitting of one or more of the following accessories which are not just refinements, but can have considerable practical application.

Sextant Mirrors

It is important when observing stars to have large mirrors to increase the field of view. Most modern sextants have these large-sized mirrors and it is a point that should be considered when purchasing sextants. A recent introduction is the wide view horizon mirror, originally developed by Davis Instruments in the US. C-Plath and Cassens & Plath offer similar mirrors and the advantage is that the horizon can be seen through the whole of the horizon glass rather than just the normal half. By coating

Chapter Seven

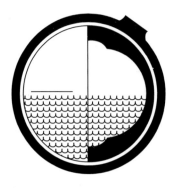

With an astigmatiser fitted, the image of a star is stretched out into a line making it easier to line up with the horizon.

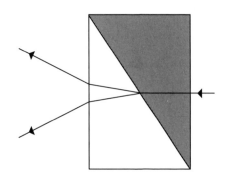

The Wollaston Prism showing how the single ray of light divides to form two images.

the horizon glass with a coating of minute quartz crystals, the entire surface of the mirror becomes reflective but is still transparent. It is suggested that the horizon is not seen so clearly though this type of mirror, but in practice there is little difference and this type of mirror does make it easier to bring stars and the sun down to the horizon. However, it does lack the vertical reference of the division between the two halves which is found with traditional split mirrors and the observer has to be more careful to check that the sextant is upright at the moment of taking the observation.

Astigmatisers And The Woollaston Prism

Astigmatisers expand the image of a star from the normal pin prick of light into a horizontal line. The theory is that the line is easier to line up with the horizon than the dot of light. In using such a device it should be remembered that the horizontal line will not necessarily indicate whether the sextant is vertical or not and the normal swinging procedure still has to be used to check this.

The Wollaston Prism is of considerable practical value when taking star observations, especially with a somewhat hazy or cloudy horizon. By its use the star image is split into two bright reflections of the star, separated by about 1/4 of a degree, one above and one below but equidistant from the star's original position. The prism is mounted on the sextant in a position as one of the index shades and when observing a star this prism is turned in like a shade.

When taking an observation of a star with the Wollaston prism, the observer of course, sees two stars, one above the other and the horizon is brought dead between the two stars' images so that perfectly correct altitudes are obtained. Accurate observations may be obtained by its use even when the horizon beneath the star is not well defined. An alternative to the Wollaston prism is the double star prism which gives a similar image but is less expensive.

Nicol Prism And Polarisers

The purpose of these items is to eliminate horizon glare and to help in giving a clear definition of the horizon. They are fitted to the telescope like an eye shade. The polarising shade offers the option of being fitted as a telescope eye piece or as a replacement for one of the standard horizon shades. Polarising shades allow more light through for the same degree of protection from the bright sun and on many of these shades the

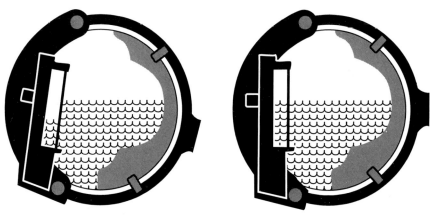

The Davis Prism Level indicates when the sextant is not level as shown in the lefthand diagram and when it is level as shown in the righthand diagram.

strength of glare protection can be adjusted by rotating one part of the shade in reference to the other.

Prism Level

We have already said how important it is to have the sextant upright when taking a sight. An experienced user can judge this by swinging the sextant about the vertical but some sextants can be fitted with a prism level which gives a positive indication that the sextant is vertical. The prism level clips on to the horizon mirror and the prism shows a refracted section of the horizon against the real horizon. If the sextant is upright, the two horizons are in line and any tilt of the sextant can clearly be seen by the displacement of the refracted horizon in the prism.

With the many attachments available, there is a distinct possibility that a sextant could become something like a Christmas tree with its many attachments. Before spending money on them, try and evaluate their worth. A practised observer will get adequate sights without their use but of course, sights can always be made easier and perhaps more reliable with the aid of the accessories. Only experience will really show their worth.

CHAPTER EIGHT

Practical Notes On The Care Of The Sextant

The sextant is both strong and fragile. The frame is built to be as rigid as possible in order to avoid distortion and the mirror frames and other fittings are made as strong and stiff as is practical but because of the very fine accuracy to which the sextant can measure it is, of necessity a fragile instrument. Small knocks and bangs can spell death to the accuracy of a sextant and so the first rule in the care of the sextant is to avoid any chance of the sextant coming into contact with other objects. The sextant is the navigator's best friend at sea so always treat it with the greatest respect. Try not to lend it to others because then you will know exactly the sort of treatment it has had. Store the sextant away from damp and heat and locate the sextant box securely so that it is not subject to vibration or jolting.

This careful handling of the sextant is relatively easy to achieve on board a ship. The motion of a ship is fairly predictable and you can comfortably have one hand for the ship and one for the sextant. The main risk on a ship comes from putting the sextant down casually after you have taken the time and are noting it down. A sudden lurch could sent the sextant flying with disastrous results. Try and get into the habit of putting the sextant back in its box or at least put it down somewhere soft like a settee where it is less likely to slide.

On a yacht the risks of accidental knocks is much higher. One free hand is rarely enough to move about a yacht safely so either pass the sextant over to another person before moving or take your sights from the relative security of the cockpit. A neck strap for the sextant at least keeps it from being lost if you have to make a sudden grab with both hands, but it is unlikely to prevent damage. For the single-handed sailor it is important to develop a routine of sight taking which seeks to minimise the risk of knocking the sextant.

Chapter Eight

Tips On Using The Sextant

When holding the sextant for a sight a firmer grip can often be achieved by using a finger or two to balance against the frame as well as holding the handle. On a vernier sextant the clamp screw should not be forced down hard when clamping the index bar, but just tight enough to prevent the index bar slipping so that the tangent screw will act. This will avoid strains and possible distortion and the same goes for the tangent screw which should not be constantly allowed to get to the end of the thread.

Always rub over your sextant lightly after use with a piece of chamois leather which is usually kept in the sextant box for this purpose. The chamois leather will absorb any damp and remove this gently from the sextant and thus prevent corrosion starting on any exposed metal. The silvering on the mirrors can be sensitive to damp, particularly on older sextants and should be wiped with great care, checking for index error after doing so. A soft tissue dipped in alcohol can be a good way to clean the mirrors. The silvering on modern sextant mirrors tends to be much more robust and less sensitive to damp. Before putting the sextant away also rub the arc dry and check that there is no damp on the thread grooves under the arc.

When taking sights in rough weather salt spray can coat the sextant. Light spray can be removed as above but it can be a good idea to wash out the chamois leather in clean fresh water afterwards to remove any salt contamination. Salt attracts damp and unless all salt is carefully removed from the sextant and especially the mirrors then deterioration is likely to set in. On a small craft there can be a very real risk of the sextant getting drenched in salt water, particularly if you are trying to snatch a sight after several days of overcast and you have to take risks to establish the position. Wiping is not going to remove this level of salt water effectively and one solution here is to use a soft brush and fresh water to clean the salt water off. Another possibility is to immerse the sextant in clean fresh water after first removing the telescope and any battery and light. It may sound extreme, but the sextant is wet anyway and it is vital that the salt is removed as quickly as possible. After removing the salt, wipe the sextant as dry as possible and then dry completely over a very

Practical Notes On The Care Of The Sextant

gentle heat. A particularly sensitive area to water contamination on the sextant is the dividing line between the plain glass and the silvered glass on the horizon mirror but be careful not to apply excessive pressure on the mirrors or their adjustments when cleaning and drying or the adjustment may be upset.

A sextant should never be left exposed to the sun for longer than necessary because it can heat up rapidly and possibly distort. Between sights it is very advisable to stow it away in its box to reduce the chance of a knock or excess exposure to the heat of the sun.

When using a micrometer sextant firm pressure should be applied to the quick release clamp so that it disengages and engages cleanly. There is a risk, particularly when engaging the clamp and the micrometer thread with the thread on the arc of dragging the micrometer thread over the arc thread with consequent damage. Be sure that you have stopped moving the index arm before releasing the clamp. The arc and micrometer teeth must be kept clean in order to prevent any stiffness occurring in the movement of the index bar. There are two schools of thought about oiling the micrometer thread, one suggesting that it gives a smoother operation of the micrometer and reduces wear on these critical threads, the other suggesting that any oil will attract dirt and dust to stick which will add to the wear and make the operation less smooth. Wear on the arc thread is a minimal problem if it is left dry so on balance it is suggested that this area should not be oiled. In the past it has always been suggested that the working parts of the sextant should be lubricated with the special oil provided in a small bottle located in the sextant box. This oil is omitted from modern sextant boxes which indicates that its use is not necessary. If oil is used on an older sextant it should be used sparingly on the underside of the large micrometer pivot screw, the micrometer screw bearing and also to the rubbing surfaces of the arc.

Routine Checking

In addition to adjusting the sextant and finding the index error, a periodical examination should be made as follows:–

Chapter Eight

■ See that the index bar moves freely along the arc without any sign of hesitation. If it does appear to bind and the sextant is clean then it is time to send it for overhaul.

■ Holding the sextant, arc away, as for the first adjustment, see if the true and reflected horizons are in line, if not note the difference. Now move the index bar round the arc from one side to the other and see if there is any variation in the distance between the true and reflected arcs. If there is, the index mirror must be distorted and again it is time to send the instrument for overhaul.

■ Examine the sextant and see if the telescope is parallel with the arc. If it appears to be out of line then this is a fault which can only be corrected by the maker or a reputable repairer.

■ The vernier should be examined to see whether, when the zero exactly cuts a division on the arc, the 10' division on the vernier also coincides with another division on the arc. It should do this at several different places along the arc, otherwise the sextant should be sent for overhaul.

An occasional rub up of the arc with the bare finger will sharpen up the marks and enable the reading to be made more easily. Be careful not to rub the arc with anything abrasive which will cut into the metal of the arc. On a vernier sextant it may sometimes be necessary to rub the arc gently with a little lamp black and thin oil and then wipe it off. This, in addition to generally cleaning the arc, makes it much more legible. If the sextant is

Photograph: Peter Chesworth

Taking sights is a two-handed job so you need to be well secured on a yacht to take sights safely.

being stowed away for a long period, the arc should be coated lightly with vaseline.

Adjusting The Sextant

Never adjust a sextant too much. If there is an index error it can be better to leave it and apply the correction to the reading rather than make the adjustment under what may be less than ideal conditions. If you make the correction on a calm night when the vessel is at anchor then the correct adjustments can be found by a star observation. Once adjusted you can then let the adjusting screws set well, and the chances are that the sextant will then stay correct for some time. Use the highest power of the inverting telescope for the final adjustment of index error and side error.

When the index error has been corrected as far as possible any small remaining correction should, if possible, be "off the arc". This will then make it a plus correction which is usually simpler to apply than a minus one. Never adjust your sextant by using a near object because owing to parallax the results will be unreliable. When using a star to adjust the sextant, do not use one which is too bright. It is preferable to use a star of the second magnitude which will give a sharper image.

Buying A Sextant

When buying a new sextant a reputable maker's name should be sufficient guarantee of the quality and accuracy of the original construction. Do not worry too much about dirt and corrosion if they do not affect the moving parts. A sextant can easily be cleaned whereas any distortion or flaw could give years of trouble and may not be possible to rectify. In general, check a second hand sextant to see that the shades and mirrors are located securely and not loose. A vernier scale can be checked as above and on a micrometer sextant both arc and micrometer screw should be clean and free from corrosion or visible defect. Reverse the instrument and see if the true and reflected arcs are in the same continuous line. Note whether the arc is well cut, as this is often worn by constant polishing. See that all the adjustment screws are

Chapter Eight

workable and not seized with corrosion or dirt. The silvering on the mirrors may be defective but this is not a serious fault as it is relatively simple to have them re-silvered. Check the telescope or telescopes for clarity and that they are of suitable power and field of view for your requirement. Before getting too excited about a cheap secondhand sextant do ask yourself why it appears to be such a bargain and if neglected and dirty, why it has been allowed to get into that state.

CHAPTER NINE

The Sextant & Coastal Navigation

The use of the sextant for fixing the position during coastal navigation appears to have fallen into disuse partly because the sextant tends to be regarded as a deep sea instrument and partly because few dedicated coastal vessels carry a sextant. Even when a sextant is carried a three point fix by horizontal angle requires the use of a station pointer as well as the sextant and finding a large enough chart table on which to use a station pointer is not easy in a small vessel or a yacht. Even if there isn't space for station pointers, the vertical sextant angle can still give valuable information for navigation.

By not using the sextant in this way navigators are missing out on very valuable navigation information. Measuring the angular height of a lighthouse or a mountain (or any charted object where the height is known) provides a distance off the object. This distance off gives a position line in the most valuable direction, parallel to the shore and it can be used as a clearance distance when there are off-lying rocks, often at a time when the conventional bearing fix with the compass cannot be used because there are insufficient objects in view.

Measuring A Vertical Angle

Set the index bar at zero and use the star telescope or simply the blank tube. Look directly at the object, and by moving the tangent screw or micrometer wheel bring the reflection of the upper part of the object which is seen in the silvered half of the horizon glass, down to the level of the lower part of the object which is seen through the plain glass in the horizon mirror.

The height of an object is usually found from information on the chart. It should be borne in mind that all heights given on the chart are heights above high water and so strictly speaking the angle measured by

Chapter Nine

the sextant should be that from the top of the object to the high water mark. It is often possible to identify the high water mark on the rocks but it is not critical to find this mark as any error involved in using the water line instead of the high water mark will tend to place the vessel closer to the shore so there is a safety margin built in. When observing the sextant angle of a lighthouse, it must be borne in mind that it is the centre of the lighthouse light that should be reflected down to the sea as it is the height of the light which is given on the chart, although the List of Lights will generally give the height of the tower itself if you find it is easier to use the top of the tower.

When measuring the distance off by vertical sextant angle in this way, the object should be close to or at the water's edge. Lighthouses are an obvious object to use, but the highest point of a cliff or headland could be equally suitable providing its height is known. Mountains would appear to be a useful object for angle measurement, but as they are usually someway inshore from the coast, the angle measured from the top of the mountain to the waterline will tend to be larger than if it were measured from a waterline directly below the object. This larger angle would tend to indicate that you are closer in than you actually are so once again you are on the safe side but this type of vertical angle using objects inland could lead to fairly large errors.

As very small angles are being observed, and as accuracy is usually required take a double reading of the angle, one taking the top of the object down to sea level and the other taking the sea level up to the top. In this way you will have one reading on the arc and one off the arc. Take the mean of the two readings by adding them and dividing by two. This will not only serve as a check on the readings, but it will also eliminate any sextant index error.

Position By Vertical Sextant Angle

A single sextant angle of a lighthouse or other object will give a position line which can be very useful as a clearing line for rocks or other dangers lying off the headland. If at the same time the bearing of the object is taken by compass, the vessel's position can be fixed with considerable accuracy. The vertical sextant angle is not subject to any errors such as deviation or variation or the tidal errors which can affect the accuracy of a bearing or running fix. However it should be noted that a vertical sextant angle fix of this type can only be used in daylight because you need to see the waterline to obtain the angle.

The position can be fixed with considerable accuracy by using two or more vertical sextant angles provided that the objects bear at least 45 degrees from each other so that the resulting position lines (curves) have a good cross. When you have obtained the vertical angle by measurement it is converted into distance by using tables found in most nautical almanacs and many navigation books. If tables are not available, then the distance off can be calculated by trigonometry or there is a rough method of calculating as follows:–

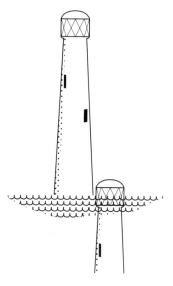

Taking a vertical angle of a lighthouse showing the true and reflected images seen in the horizon mirror.

■ Multiply the height by 1.856 and divide by the sextant angle in minutes to get the distance off.

■ Up to five miles distance this will be approximately correct, and as the height can be multiplied by the constant beforehand, it is very simple to divide by the angle so the position is obtained in a few seconds.

The Vertical Danger Angle

We have already seen how a vertical sextant angle can be used to give a clearing line to keep clear of off-lying dangers. Rather than plot the line continuously on the chart it is possible to set the sextant to an angle

Chapter Nine

which will give instant warning whether the vessel is inside or outside the required line. Look up the angle for the required distance off by entering the tables with the appropriate height. This is the angle to which the sextant must be set. The vertical angle and hence the distance off will vary as you approach the object, so if the sextant angle becomes greater than the danger angle previously set on the sextant then the vessel is too close in and the course needs to be altered out. If the sextant angle is smaller than the danger angle, then the vessel is safely outside the danger circle indicated by the danger angle on the sextant. This can be remembered easily because the nearer you approach an object, the larger the angle between the observer's eye and the top of the object must be.

The Horizontal Sextant Angle

Three objects are required that are marked on the chart and which are clearly identifiable. The horizontal angle is measured between the middle one and that on either side of it using the sextant and this will give a fix which is independent of the compass and deviation, a valuable source of information if the compass is suspect. The position obtained by this method is extremely accurate and this is the method used for marine survey work when electronic position fixing systems are not available and this use accounts for the special survey sextants which are available on the market. These are lightweight sextants because they may have to be used for long periods and they can be identified by the absence of shades and the large mirrors which are fitted.

To take the horizontal angle between two objects, hold the sextant horizontally ie:– with the frame and handle in the horizontal position and the mirrors upwards. Bring the reflection of the right hand object directly below the left hand object when the latter is seen through the plain part of the horizontal glass. To pick up the reflection of the right hand object, the index bar should be at zero, and the observer looks directly towards the object. If it is necessary to reflect the left hand object over the right hand object the sextant must of course be held upside down although this is rarely necessary. The horizontal angles obtained should then be set on the station pointer which is used to plot the position on the chart. Alternatively, if a station pointer is not available then the angles can be plotted onto tracing paper which can then be used to transfer them to the chart.

Care must be taken in selecting the objects to be used and ideally they should be in a straight line or with the centre object nearer the observer. If the three objects form the arc of a circle when joined they will not give a satisfactory fix. It is possible to use a horizontal sextant angle as a danger angle when coasting so that it can provide a ready check whether the vessel is too close to a danger. This is rather similar to using vertical sextant angles for the same purpose and it will largely be a case of using what objects are visible and whether their height is known. The vertical sextant angle is usually preferred for this purpose because it is simpler to identify and calculate the distance from tables.

It is possible to plot lines on the chart, each line being a position line for a particular horizontal angle. By plotting a lattice of such lines, a grid can be built up which can allow horizontal sextant angles to be plotted manually without the need for a station pointer. This is the method used by surveyors working in a set area and it could be useful for an anchorage which is used regularly.

The sextant is an under-used tool for coastal navigation. It can provide a great deal of very accurate and valuable information and it is the fact that this information is not affected by any variable errors or outside influences which makes it so valuable. With the availability of relatively cheap plastic sextants the potential of the sextant for coastal navigation is worth exploring.

*

Index

A
Accessories 5, 87
Adjusting screws/clips 15
Adjusting the sextant 95
Adjustments/errors 67
Altitudes maximum & minimum 46
– *meridian* 39
– *moon* 36
– *planet* 38
– *star* 36
Arc/limb 10
Astigmatisers 88

B
Blank tube 83
Buying a sextant 95

C
Care/using a sextant 91
– *handling* 4
– *routine checking* 93
Chronometers 47
Clamping screw 7, 13
– *vernier sextant* 61
Collimation 75
– *telescope* 84

D
Decimal graduations 66
Description of the sextant 1

E
Electric Bulb 14
Endless tangent screw sextant 8, 62
Errors/adjustments 67
Errors – *adjustable* 70
– *centering* 68
– *certificate* 70
– *collimation* 75
– *graduation* 68
– *index* 74, 78
– *non-adjustable* 68
– *perpendicularity* 70
– *shade* 68
– *side* 72

F
Frame 9

G
Galilean telescope 83

Index

H
Handling ... 4
Horizon glass ... 11
Horizon shades ... 11
Horizontal sextant angle ... 100

I
Index bar ... 12
Index glass ... 12
Index shades ... 13
Inverting telescope ... 83, 85

L
Latitude – *by planet in daytime* ... 53
Laws of optics ... 1, 19
Limb/arc ... 10

M
Magnifier ... 14
Measuring – *vertical angle* ... 97
Magnifier ... 14
Meridian ... 25
Meridian altitudes ... 39
– of moon, planet, stars ... 40
– of sun ... 39
Micrometer sextant ... 8
Micrometer tangent screw sextant ... 59
Mirrors ... 87

N
Nicol Prism ... 89
Notebook ... 49

O
Optical laws ... 1, 19

P
Parts of the sextant ... 9
Plastic sextant ... 16
Polarisers ... 89
Position line ... 57
Prismatic telescope ... 83, 86
Prism level ... 90

R
Reading the sextant ... 64
Removable parts ... 14
Rough weather sighting ... 49

S
Sextant certificate ... 3, 70

Index

S
Sight notebook 49
Sights 27, 41
– *observations of moon* 55
– *observations of pole star* 53
– *observations of stars* 52
– *observations of sun* 50
– *position lines* 57
– *prime vertical* 57
– *rough weather* 49
– *simultaneous* 55
Sight vane 83
Star telescope 83, 87

T
Tangent screw 13
Telescope alternatives 86
Telescope collar 15
Telescope shades 6
Telescope types 83
Telescopes 83
Time taking 47
Types of sextant 7

U
Using/care of sextant 6, 92

V
Vernier 13
Vernier sextant 63, 62
Vertical angle 97
Vertical danger angle 99
Vertical sextant angle 98

W
Woolaston Prism 88

Z
Zenith 23